Air Pollution Governance in East Asia

Focusing on Taiwan, South Korea, Japan, and Mainland China, the contributors to this book analyze various cases of air pollution within East Asia.

Air pollution in East Asia is a major health risk, which also has damaging impacts on the environment leading to impacts on society, economic growth, and welfare. While existing laws and policies have made progress in alleviating air pollution in each country in the region, the protection of favorable environments and the resolution of transboundary air pollution problems have become major targets of regional cooperation. Combining perspectives from social sciences and science, technology, and society studies, the contributors to this book examine both the technical and socioeconomic-political aspects of these challenges through a range of case studies from around the region.

The book is a valuable read for researchers and policymakers looking at air pollution and transboundary governance challenges within and beyond East Asia.

Kuei-Tien Chou is Professor, Graduate Institute of National Development, National Taiwan University (NTU). Director, Risk Society and Policy Research Center, NTU, Taiwan.

Koichi Hasegawa is Specially Appointed Professor, Graduate School of Comprehensive Human Sciences, Shokei Gakuin University, Japan.

Dowan Ku is Director of Environment and Society Research Institute, South Korea.

Shu-Fen Kao is Associate Professor, Department of Sociology and Social Work, Fo Guang University, Taiwan.

Routledge Contemporary Asia Series

For more information about this series, please visit: www.routledge.com/Routledge-Contemporary-Asia-Series/book-series/SE0794

Air Pollution Governance in East Asia

Edited by Kuei-Tien Chou,
Koichi Hasegawa, Dowan Ku
and Shu-Fen Kao

Routledge
Taylor & Francis Group

LONDON AND NEW YORK

First published 2022
by Routledge
4 Park Square, Milton Park, Abingdon, Oxon OX14 4RN

and by Routledge
605 Third Avenue, New York, NY 10158

Routledge is an imprint of the Taylor & Francis Group, an Informa business

British Library Cataloguing-in-Publication Data
A catalogue record for this book is available from the British Library

Library of Congress Cataloguing-in-Publication Data
A catalog record for this book has been requested

ISBN: 978-1-032-07834-2 (hbk)
ISBN: 978-1-032-07836-6 (pbk)
ISBN: 978-1-003-21174-7 (ebk)

DOI: 10.4324/9781003211747

Typeset in Galliard
by Apex CoVantage, LLC

Contents

Figures

Tables

Contributors

Jusen Asuka is Professor, Center for Northeast Asian Studies, Tohoku University, Department of the Environmental Science, Tohoku University, Japan.

Chelsea C. Chou is Associate Professor, Graduate Institute of National Development, National Taiwan University, Taiwan.

Kuei-Tien Chou is Professor, Graduate Institute of National Development, National Taiwan University (NTU). Director, Risk Society and Policy Research Center, NTU, Taiwan.

Koichi Hasegawa is Specially Appointed Professor, Graduate School of Comprehensive Human Sciences, Shokei Gakuin University, Japan.

Manami Horihata is Professor, College of Art and Science, J.F. Oberlin University, Japan.

Shih-Hao Jheng is Research Assistant, Institute of Sociology, Academia Sinica. MA. Department of Sociology, Chengchi University, Taiwan.

Paul Jobin is Associate Research Fellow, Institute of Sociology, Academia Sinica, Taiwan.

Shu-Fen Kao is Associate Professor, Department of Sociology and Social Work, Fo Guang University, Taiwan.

Inkyoung Kim is Assistant Professor of Political Science, Bridgewater State University, United States.

Minjae Kim is Researcher, Environment and Society Research Institute, South Korea.

Dowan Ku is Director of Environment and Society Research Institute, South Korea.

Kuan-chen Lee is Assistant Research Fellow, Institute for National Defense and Security Research, Taiwan.

Sanghun Lee is Professor, Peace College of Liberal Arts, Hanshin University. Director of Center for Regional Development, Hanshin University, South Korea.

Taedong Lee is Professor, Department of Political Science & International Studies. Director, Environment Energy and Human Resource Development Center, Yonsei University, South Korea.

Wooyeal Paik is Associate Professor, Department of Political Science & International Studies, Yonsei University, South Korea.

Chih-sung Teng is Associate Professor, Graduate Institute of National Development, National Taiwan University, Taiwan.

Wen-Ling Tu is Professor, Department of Public Administration, National Chengchi University, Taiwan.

David Walther is Postdoctoral Research Fellow, Risk Society and Policy Research Center, National Taiwan University, Taiwan.

Paulina PY Wong is Assistant Professor, Science Unit, Lingnan University, Hong Kong, China. Centre Fellow, Institute of Policy Studies, Lingnan University, Hong Kong, China.

Chin-en Wu is Associate Research Fellow, Institute of Political Science, Academia Sinica, Taiwan.

Chee Wei Ying is Ph.D. student, Graduate Institute of National Development, National Taiwan University, Taiwan.

1 Cosmopolitan governance to transboundary air pollution in East Asia

Shu-Fen Kao and Kuei-Tien Chou

East Asia suffered 35% of the global burden of mortality from ambient (outdoor) air pollution in 2015, a higher proportion than in any other region.[1] Air pollution in East Asia is not only a major health risk; it also has damaging impacts on the environment, which leads to significant economic and social consequences, dampening economic growth and reducing welfare. Although existing laws and policies have made progress in alleviating air pollution in individual East Asian countries, the protection of vulnerable ecosystems and human environments, together with the resolution of transboundary air pollution problems, has only recently become major targets of regional cooperation.

In common with other complex societal challenges—for example, climate change, genetically modified organisms (GMOs), chemical pollution, avian flu, or bovine spongiform encephalopathy (BSE)—air pollution was initially regarded as a simple environmental problem. Over time, however, perceptions of these issues changed, and they are now seen as hybrid challenges characterized by scientific uncertainty, invisibility, and transboundary risks. Changing perceptions have, in turn, facilitated a paradigm shift in environmental research. The shifting research paradigm has evolved to embrace a trend toward interdisciplinarity in tackling the challenges studied by environmental sociology (Chou, 2015).[2] Researchers who investigate air pollution and other environmental problems characterized by technological disputes and uncertainty increasingly emphasize interdisciplinary and transdisciplinary integration of research findings (Gross and Heinrichs, 2010). Their research is also increasingly viewed in the context of the risk society and risk governance (Sellke and Renn, 2010).

Social scientists have investigated the processes by which scientific knowledge about air pollution is produced and the contested construction of air pollution risk. Writing on the subject of the risk society, Beck (1986) argued for the emancipation of technology from science and a return to the sort of scientific autonomy envisaged by the Age of Enlightenment. Such an autonomous science would be unencumbered by the interference and deformation introduced by political decision-making. Jasanoff (1990) and Fischer (2000) also emphasized the importance of scientific knowledge in framing environmental regulations, politics and advocacy, as well as the types of scientific knowledge generated by epistemic communities. The intrusion of science into these areas has generated

DOI: 10.4324/9781003211747-1

contested discourse among environmental activists, civil society groups, and policymakers.

Lidskog et al. (2010) proposed different research approaches to the analysis of expertise and technocracy. These authors focused on how experts face up to public challenges, how scientific and local knowledge are produced, and the democratization of expertise. Nowotny (2003) described the deployment of socially robust knowledge to challenge the expertise of mainstream scientists, and Ottinger (2010) examined how civil society can produce scientific knowledge that challenges official scientific discourse or the scientific data released by polluters.

The management of air pollution is now entangled with these issues of contested expertise and competing scientific narratives, and management is further complicated by transboundary risks. On the one hand, the analysis of transboundary risks represents an emerging approach to air pollution risk governance. On the other hand, transboundary air pollution risks can no longer be investigated at the level of individual nations and need to be analyzed in the context of global or regional governance structures (Chou, 2015). *Air Pollution Governance in East Asia* interrogates the risks posed by transboundary air pollution in the context of four East Asian countries and analyzes cosmopolitan governance of such risks in the context of regional traits, political structures, and socioeconomic problems.

Bulkeley (2005) indicates that transboundary risks represent, in essence, environmental and technological issues that involve several levels of government, span multiple spatial scales, and diffuse across borders. New research methods are therefore needed to incorporate cross-disciplinary and large-scale issues of scientific uncertainty that attend the analysis and perception of issues such as air pollution and climate change in different countries and regions. In light of the tendency of nations to approach these issues idiosyncratically, there is a need to replace methodological nationalism with methodological cosmopolitanism. The vision and structure of social science research, therefore, need to move beyond traditional analyses framed by national norms and attitudes toward more cross-border approaches and a global vision (Beck and Sznaider, 2006; Beck and Grande, 2010). This major paradigm shift in the field of social science research would also have impacts on political and economic critiques of these issues, and on governance research (Chou, 2015).

The research methodology of *Air Pollution Governance in East Asia* adopts methodological cosmopolitanism to analyze transboundary air pollution risks that affect Japan, South Korea, Taiwan, and China. Within this framework, the contributing authors attempt to understand the characteristics of transboundary air pollution risk encountered by each country. In the chapters that follow, they move beyond the analysis of governance conflicts generated by internal, nation-state political and economic structures to investigate trans-national political, cultural, and geographical affinities that could form the basis of a common governance model.

There are research precedents that support our approach. Chang (1999) coined the concept of "compressed modernity" in their analysis of South Korea's exponential economic growth. This growth was based on the rapid adoption of Western models of economic development, which led to social imbalances due to the temporal compression of development and the pursuit of growth through technology. As a result, South Korea suffered a series of major economic disasters in the 1990s. Chou (2000, 2002, 2004) also adopted the concept of a "delayed, hidden high-tech risk society" to analyze the accelerated industrialization that Taiwan initiated in response to the threat of global competition. Taiwan aimed to catch up to other regional powers in terms of scientific and technological development, but the absence of scientific analysis of these developments resulted in tremendous hidden technological risks. Delays in governance and regulation only led to even greater social and political risks.

Han and Shim (2010) used the working hypothesis that techno-industrial development pathways adopted by East Asian countries led to them being vulnerable to a common set of socioeconomic and environmental risks. After analyzing large-scale risk and disaster experiences among East Asian countries, Han and Shim concluded that risks may therefore be "regional." Furthermore, if transboundary air pollution is one of the deficiencies caused by the rush to modernization (Beck and Grande, 2010), it becomes worthwhile to adopt a perspective of "embedding the regional" in order to explore the transboundary characteristics of air pollutions. In particular, transboundary air pollution could be used as a template for the development of cosmopolitan governance among East Asian countries.

Technological elites and authoritarian technocracies in Taiwan, Japan, Korea, and China have dominated science and technological policy decision-making to put their countries on fast-track industrialization. These attributes have led East Asian countries to imitate the techno-industrial developmental models that led to the industrial and, later, socioeconomic modernization of the West.[3] This rush for modernization led to relaxed regulations and a laissez-faire approach to technological risks (Chou, 2015).

The literature reviewed above leads to the conclusion that transboundary air pollution is inevitably a cosmopolitan issue (as constructed by Beck, 2002). Furthermore, controversies centered upon the use of technology and its impacts on nation-states have resulted in what a number of researchers refer to as cosmopolitan risk communities (Beck, 1996, 2009; Zhang, 2015). Regardless of whether the risks center on GMOs, BSE, nuclear accidents, or air pollution, these issues constitute transboundary and cross-border threats, and in this regard, the fates of South Korea, Japan, Taiwan, and China are tightly intertwined. In other words, even if these societies already had a certain degree of cosmopolitanization (Beck and Levy, 2013), the risks would have forced them to develop it further. Our goal is to try to understand the extent to which these East Asian societies have experienced cosmopolitanization to date and to investigate the possibility that they have already produced a cosmopolitanization risk collective (Beck and Levy, 2013). We also investigate whether the existence of risk communities among East Asian states could foster the emergence of transnational actors, activities, networks, institutions, or standards in the government or civil society (Grande, 2006). Otherwise, they should be seen as latent cosmopolitan risk communities independent of one another, having to deal with the pressures and regulations of hidden cosmopolitan risk governance in their individual countries. If this were the case, would it result in the isolation, fragmentation, and fragility of risk management within individual countries, with the consequence that their respective governances and civil societies would lack the means to develop truly international cooperation and governance of transboundary issues?

The structures of risk governance that are unique to East Asia must be understood from the perspective of contemporary technocracy and regulatory science. Based on the operational experiences of regulatory science in western industrial countries, Jasanoff (2005) believed that contemporary technological affairs are dominated by technological bureaucracies. Such bureaucracies exert an invisible, sometimes monopolized domination over technological affairs in some countries, a state of affairs that disrupts democratic decision-making and generates substantial disputes. Facing all types of technological risks, modern societies have gradually evolved from being passive victims of technology to being able to provide mature reflections upon and critiques of alternative pathways toward sustainable development (Nowotny et al., 2001). In particular, some societies have developed robust responses to technologically mediated impacts on the environment, ethics, and health, and have deployed socially robust knowledge (SRK) to monitor and challenge questionable actions by governments (Jasanoff, 2003; Nowotny and Leroy, 2009; Delvenne, 2010). During this process, if citizens can break away from their passivity and systematically develop their risk knowledge, then they may grasp the opportunity to break through the monopoly of authoritative politics to shape a technological democracy.

Using the cosmopolitan theoretical framework, *Air Pollution Governance in East Asia* examines air pollution and risk governance in four East Asian countries: Japan, Taiwan, South Korea, and China. Contributors to this edited work employ perspectives derived from interdisciplinary social sciences, particularly environmental sociology, political science, and STS (Science, Technology and Society) to analyze cases of air pollution in this region. "Air pollution" is not merely a technical problem. It is embedded in complex social-political-economic structures within and among countries. On the one hand, the four East Asian countries considered in this volume have long histories of State Developmentalism, are largely dependent on the brown economy system for growth, have traditions of authoritarian expert politics. These characteristics have delayed meaningful governance of air pollution. On the other hand, because air pollution is a transboundary issue and systemic risk, it requires cosmopolitan risk governance among East Asian countries. Given the fact that governance in these countries takes place within nation-specific political and economic contexts, there exists a dilemma of how cosmopolitan governance can be achieved. Furthermore, scientific uncertainty over the attribution of pollution resulted in China being scapegoated and allowed the other three countries to shrink away from reflecting on their own domestic sources of air pollution from manufacturing, energy-intensive industries, and automobiles.

To sum up, the objectives of *Air Pollution Governance in East Asia* are as follows: First, we plan to conceptualize and construct East Asian perspectives on air pollution governance by exploring various cases studies in this region, as well as discussing how the existing fossil fuel-based economy can evolve toward a more sustainable, less carbon-intensive one. Second, we aim to characterize the politicization of air pollution in each country and examine how transnational initiatives to monitor air pollutants are hindered by diplomatic tensions over security and other issues between Taiwan, Japan, South Korea, and China. Third, in addition to identifying the victims of air pollution at different scales (e.g., individual, neighborhood, national, international), the macrostructures of the polluters and pollution will also be investigated. Fourth, via case studies, contributors to this volume demonstrate the contested construction of air pollution risk controversies and discuss how the framing, scientific methodologies, and conflicts of interests between different camps hamper risk governance. Finally, we seek to provide suggestions for better governance models for the risks posed by air pollution, using lessons learned from East Asian experiences.

Following the introductory chapter, the book is structured into four parts: "Air Pollution Politics in East Asia," "Regional and Transboundary Air Politics," "National Air Pollution Battles," and "Contested Risk Constructions of Air Pollution." Part I deals with the politicization of air pollution in Korea and Taiwan. In Chapter 2, Kim and Ku's chapter explores how particulate matter (PM) has become a politicized issue in South Korea. The authors illustrate specific aspects of contestation between social forces on the PM issue, and they go on to argue that prevailing populistic responses to PM reflect a failure of environmental politics. This chapter analyzes air pollution politics by drawing on research into the relationship between the capitalist state and environmental imperatives. The authors conclude that in South Korea, policy innovation has been limited by its export-oriented fossil-fuel-based capitalist economy with its established connections between policy elites and industries.

In Chapter 3, Lee proposes an explanation rooted in political ecology for why PMs have received greater attention and become more politicized than other air pollutants. Using "state-nature" theory and "methodological cosmopolitanism," Lee explains that PMs were politicized by a state-led "framing strategy" (composed of centralization and territorialization strategies). The result, according to Lee, was that fundamental countermeasures against PMs were established at a far from satisfactory level. Furthermore, it is difficult to come up with solutions to PM pollution via cooperation with neighboring countries. Lee argues that a cosmopolitan framing strategy must be developed that resists inadequate state-level framing strategies. Additionally, Lee believes that inter-urban solidarity against PMs must be encouraged among East Asian states before the fundamental root causes of PM pollution can be tackled.

Part II consists of three case studies, each of which deals with regional transboundary air politics. In Chapter 4, Jobin and his colleagues introduced the notions of "air politics" and "air diplomacy" to address the issue of transboundary air pollutants blowing from China to Taiwan, and their influence on Taiwan's domestic politics. The authors first describe academic discussions on air pollution that have taken place between Taiwanese and Chinese scholars. Despite the radical asymmetry of power between the two sides of the Formosa strait, these meetings have been conducted smoothly for a decade. But transboundary air pollutants remain a taboo within these discussions. Given the crucial importance of China to Taiwan's domestic politics, the China factor should logically play an important role in domestic air politics. Nevertheless, as the authors show in the second part of the chapter, domestic anti-pollution activists have emphasized the air pollution burden imposed by heavy industries in south and central Taiwan. Although air pollution has been a significant concern, in particular during local elections, Jobin and colleagues found that transboundary pollutants from China played only a marginal role in the outcome of the 2018 local and municipal elections.

Chapters 5 and 6 explore power asymmetries among countries in the East Asian region, and how such asymmetries are interwoven with the challenges of transboundary air pollution governance. In Chapter 5, Lee and Paik illustrate how asymmetric power barriers have seriously hampered negotiations between China and South Korea. They introduce the concept of "atmospheric politics" to study efforts by governments, regional epistemic communities, and policymakers to frame transboundary air pollution governance in the context of transnational and multilateral frameworks of cooperation or conflict resolution. Asuka, in Chapter 6, documents the historical development of air pollution governance in Japan. As in Lee and Paik's chapter, Asuka emphasizes how transnational initiatives to monitor air pollutants in East Asia have been hindered by diplomatic tensions between Japan, South Korea, and China centered on issues unrelated to air pollution.

Part III engages in the analysis of air pollution policy battles. Chapter 7 by Chou and Walther reveals that air quality and improvements in Taiwan have been unequally distributed. Highly polluted areas in Taiwan have experienced less improvement in air quality than those with less ambient air pollution. The areas with high PM2.5 pollution are more likely to be in low-income agricultural counties and cities, and these areas overlap with zones of low education and high crude death rates. Chou and Walther propose that Taiwanese government policies should be targeted and differentiated in accordance with local social and economic conditions to achieve a just and more equitable transition to improved air quality. They conclude that a philosophy of "just transition" can help to uncover key issues in developmentalism and injustices embedded in Taiwan's neoliberalism. An emphasis on just transition may also force the Taiwanese government to confront the inequality perpetuated by the neoliberal model of economic development. The authors emphasize that policies and actions to secure a just transition may constitute the next phase of environmental governance that post-developmental states need to undertake.

In Chapter 8, Chou and her colleagues analyze China's 2017 "Coal-to-Gas Switch" campaign. They find that despite its eco-friendly aspirations, China's authoritarian government has not always made sound and sustainable policies. The major goal of the step Coal-to-Gas Switch campaign was to decommission all small coal-fired boilers and to replace them with natural gas in the Beijing–Tianjin–Hebei region. To force individuals' cooperation, the Chinese government formulated a discourse focused on the presumed responsibility of individual owners to abandon coal-fired boilers. However, the campaign was only able to achieve short-term successes in achieving its blue-sky vision. By pinning its hopes for reducing PM2.5 concentrations on the dumping coal burners, the Coal-to-Gas-Switch campaign led to insufficient heating for many households in the northern provinces. Unusually cold winter temperatures then forced Beijing to slow down the campaign. The authors conclude that although the Chinese government has made up its mind to improve air quality, individualizing responsibility for air pollution abatement has led to unsustainable policies.

In Chapter 9, Horihata employs a social-history approach to analyze the Pollution Health Damage Compensation Law enacted in Japan. Using data from interviews and questionnaires, together with an analysis of the Tokyo Air Pollution Litigation, which was itself a response to the Yokkaichi Pollution Lawsuit, Horihata explored how different policy camps attempted to cope with the changing nature of pollution risks. Among other things, the source of Tokyo's air pollution has changed from factory soot to automobile exhaust and compensation for pollution-associated medical expenses has been discontinued. The Tokyo Air Pollution Lawsuit, which demanded clarification of the responsibility of automobile manufacturers and compensation for medical expenses, was settled, but the political response thereafter was ineffective, and the manufacturers' responsibility for the pollution was obscured. This case study confirms that the politics of risk in Japan emphasize the non-decline of economic activities over the health of citizens.

Part IV, "Contested Risk Constructions of Air Pollution," includes case studies from South Korea, Taiwan, and Hong Kong. In Chapter 10, Kim points out that the Korean government has pursued air-pollution-control strategies that have varied from assigning individual responsibility to seeking international cooperation. In addition to developing international cooperation on transboundary air pollution and collaboration with industries to reduce domestic sources of particulate matter, the Korean government has promoted individual responsibility for reducing particulate matter emissions among its citizens. This individualized responsibility has reaped some reduction of particulate matter emissions through emergency measures, such as restricting the operation of old diesel vehicles, and by encouraging environmentally friendly behaviors among citizens.

Kim argues that because of an institutional culture that assigns primary responsibility for air pollution mitigation to the Ministry of Environment for Korea (MOEK), the role of the private sector in reducing particulate matter remains unclear and nascent at best. Most measures proposed for industry have been recommendations without regulatory teeth. By contrast, state-owned power plants and individually operated diesel vehicles have been tightly monitored and regulated. Kim concludes that if it is to tackle particulate matter pollution successfully, the Korean government must pay much more attention to carbon-intensive industries, which is the country's highest energy user and biggest emitter of air pollutants.

In Chapter 11, Tu analyzes the role played by scientific research in political controversies over pollution and concludes that more scientific studies did not necessarily contribute to solutions in Taiwan. Polluters may exploit scientific uncertainty to manipulate and obstruct policy decisions. Tu argues that the effective incorporation of scientific knowledge into policy formulation demands a careful examination of how questions are framed, methodologies deigned, research applied, and conflicts of interests in scientific knowledge generation resolved.

In Chapter 12, Wong proposes that the use of regulations and public health advisories to manage Hong Kong's air pollution requires the rethinking and improved identification of the sources of air pollution. She presents a grounded and neutral overview of the double exposure Hong Kong residents to trans-boundary air pollution and local emissions. Issues such as the increase of local emissions, relaxation of regulations, non-compliance with WHO standards, and future challenges all demand that regulations and policies be rethought. Finally, Wong provides recommendations for improved assessment methods and adaptive solutions that could increase the effectiveness of air pollution controls. She also proposes measures to formulate better advice to the government, with the ultimate goal of making government more responsive to local air pollution problems, and to encourage preventative and mitigative actions to maximize the quality of life.

Air Pollution Governance in East Asia views air pollution through an inter-disciplinary lens, incorporating perspectives derived from political science, environmental sociology, geography, and economics. Subdisciplines covered across these domains include political and science communications, political economy, environmental, health, and economic policies, international relations, economic development, and law. This interdisciplinary breadth prevents *Air Pollution Governance in East Asia* from fitting neatly into the boundaries of a single discipline. One could argue, though, that this is a consequence of the complexity of the air pollution challenge.

As discussed by Lee and Paik (2020, and Chapter 5 of this book), asymmetric power relations remain a serious problem among East Asian countries. This asymmetry is not just seen in power relationships among governments but in the type and degree of domestic pressure exerted by citizens. For example, South Koreans may urge their government to settle with China over the problem of air pollution (including transboundary PM) while Chinese citizens complain about domestic pollution in China but follow Chinese state media in arguing that South Korea generates its own pollution (Lee and Paik, 2020: 133). The difference in political regimes—authoritarian in the case of China but democratic in the case of South Korea, Taiwan, and Japan—contributes to the nurturing of different popular reactions and expressions of green nationalism. For these and other reasons, it is imperative to improve our understanding of dynamic interactions between contested constructions of air-pollution risk within and between East Asian countries. Such improved understanding is needed for the successful resolution of transboundary air pollution problems in this region. *Air Pollution Governance in East Asia* aims to broaden this understanding.

To solve its transboundary problems, it is critical in East Asia to establish a robust epistemic community that nurtures a common understanding of the scientific complexity of transboundary air pollution. This community should provide recommendations for improved regional governance of transboundary pollution that are based on scientific knowledge and expertise. Additionally, because asymmetric power relationships hinder cooperation, more egalitarian relationships among East Asian must be established before they can successfully tackle transboundary air pollution problems together. Achieving such cosmopolitan, egalitarian, and cooperative governance will be challenging. However, the contributors to *Air Pollution Governance in East Asia* are attempting to establish a basis for cosmopolitan cooperation and to understand the barriers to such cooperation. We believe that the new empirical findings offered in the chapters of this book will provide readers with fresh understanding of the complexities of transboundary air pollution in East Asia.

Notes

1 Source: www.ccacoalition.org/en/content/air-pollution-measures-asia-and-pacific (2021.01.13 access).
2 As Chou (2015) pointed out that there was a strong body of research that touched on this development, such as works by Yearley (2010), Levidow (2001), Reusswig (2010) research on the impact of acid rain and climate change on environmental sociology, and Wynne and Dressel's (2001) analysis of the transboundary risks of BSE. Tindall (1995), Schrecker (1995), and Gross and Heinrichs (2010) also discussed the interdisciplinary trends and challenges in environmental sociology.
3 However, beginning in 2000, authoritarian expert politics under the increasing trend toward technological democracy, has been faced with fierce challenges from an increasingly robust civil society (Chou, 2009). Whether it be the compressed modernity or stagnation resulting from the rush for modernization (Chang, 2010), or the hidden and delayed technological risk society (Chou, 2000, 2009), both explanations point to the antagonism between the government and civil society; and the reality is that over the long term, a more fragile risk and regulatory culture have developed in East Asia, as compared to western industrialized societies (Chou, 2015).

References

Beck, Ulrich. (1986). Risikogesellschaft. In U. Beck (Ed.), *Auf dem Weg in einen andere Moderne*. Frankfurt: Suhrkamp.
Beck, Ulrich. (1996). World Risk Society as Cosmopolitan Society? Ecological Questions in a Framework of Manufactured Uncertainties. *Theory, Culture & Society* 13 (4): 1–32.

Beck, Ulrich. (2002). The Cosmopolitan Society and Its Enemies. *Theory, Culture & Society* 19 (1–2): 17–44.

Beck, Ulrich. (2009). Critical Theory of World Risk Society: A Cosmopolitan Vision. *Constellations* 16 (1): 3–22.

Beck, Ulrich and Grande, Edgar. (2010). Varieties of Second Modernity: The Cosmopolitan Turn in Social and Political Theory and Research. *British Journal of Sociology* 61 (3): 409–443. London: LSE.

Beck, Ulrich and Levy, Daniel. (2013). Cosmopolitanized Nations: Re-imagining Collectivity in World Risk Society. *Theory, Culture & Society* 30 (2): 3–31.

Beck, Ulrich and Sznaider, Natan. (2006). Unpacking Cosmopolitanism for the Social Sciences: A Research Agenda. *British Journal of Sociology* 57 (1): 1–23.

Bulkeley, Harriet. (2005). Reconfiguring Environmental Governance: Towards a Politics of Scales and Networks. *Political Geography* 24 (8): 875–902.

Chang, Kyung-Sup. (1999). Compressed Modernity and its Discontents: South Korean Society in Transition. *Economy and Society* 281: 30–55.

Chang, Kyung-Sup. (2010). The Second Modern Condition? Compressed Modernity as Internalized Reflexive Cosmopolitization. *The British Journal of Sociology* 61 (3): 444–464.

Chou, Kuei-tien. (2000). Bio-industry and Social Risk—Delayed High-tech Risk Society. *Taiwan: A Radical Quarterly in Social Studies* 39: 239–283.

Chou, Kuei-tien. (2002). The Theoretical and Practical Gap of Glocalizational Risk Delayed High-tech Risk Society. *Taiwan: A Radical Quarterly in Social Studies* 45: 69–122.

Chou, Kuei-tien. (2004). Monopolistic Scientific Rationality and Submerged Ecological and Social Rationality—A Discussion of Risk Culture Between Local Public, Scientists, and the State. *Taiwan: A Radical Quarterly in Social Studies* 56: 1–63.

Chou, Kuei-tien. (2009). Reflexive Risk Governance in Newly Industrialized Countries. *Development and Society* 43 (1): 57–90.

Chou, Kuei-tien. (2015). *Cosmopolitan Approach of Trans-boundary Risk Governance in East Asia*. Singapore: World Congress of Risk Analysis.

Delvenne, Pierre. (2010). Parliamentary Technology Assessment Institutions as Indications of Reflexive Modernization. In *Society for Social Studies of Science Annual Meeting with JSSTS*. Tokyo: University of Tokyo Press.

Fischer, Frank. (2000). *Citizens, Experts, and the Environment: The Politics of Local Knowledge*. Durham, NC: Duke University Press.

Grande, Edgar. (2006). Cosmopolitan Political Science. *The British Journal of Sociology* 57 (1): 87–111.

Gross, Matthias and Heinrichs, Harald. (2010). Moving Ahead: Environmental Sociology's Contribution to Inter-and Transdisciplinary Research. In M. Gross and H. Heinrichs (Eds.), *Environmental Sociology: European Perspectives and Interdisciplinary Challenges* (pp. 347–351). Dordrecht, Heidelberg, London, and New York: Springer.

Han, Sang-Jin and Shim, Young-Hee. (2010). Redefining Second Modernity for East Asia: A Critical Assessment. *British Journal of Sociology* 61 (3): 465–489.

Jasanoff, Sheila. (1990). *The Fifth Branch: Science Adviser as Policymakers*. Cambridge, MA: Harvard University Press.

Jasanoff, Sheila. (2003). Technologies of Humility: Citizen Participation in Governing Science. *Minerva* 41 (3): 223–244.

Jasanoff, Sheila. (2005). Judgment under Siege: The Three-body Problem of Expert Legitimacy. In P. Weingart and S. Maasen (Eds.), *Democratization of Expertise? Exploring Novel Forms of Scientific Advice in Political Decision-making*, Sociology of the Sciences Yearbook (pp. 209–224). Dordrecht: Kluwer.

Lee, Taedong and Paik, Wooyeal. (2020). Asymmetric Barriers in Atmospheric Politics of Transboundary Air Pollution: A Case of Particulate Matter (PM) Cooperation between China and South Korea. *International Environment Agreements* 20: 123–140.

Levidow, Les. (2001). Genetically Modified Crops: What Transboundary Harmonization in Europe? In J. Linnerooth-Bayer, R. E. Lofstedt, and G. Sjostedt (Eds.), *Transboundary Risk Management* (pp 59–90). London: Earthscan.

Lidskog, Rolf, Soneryd, Linda, and Uggla, Ylva. (2010). *Transboundary Risk Governance*. London: Earthscan.

Nowotny, Helga. (2003). Democratizing Expertise and Socially Robust Knowledge. *Science and Public Policy* 30 (3): 151–156.

Nowotny, Helga and Leroy, Pieter. (2009). Helga Nowotny: An Itinerary between Sociology of Knowledge and Public Debate-Interview by Pieter Leroy. *Natures Sciences Sociétés* 17 (1): 57–64.

Nowotny, Helga, Scott, Peter B., and Gibbons, Michael T. (2001). The Co-evolution of Society and Science. In *Re-thinking Science: Knowledge and the Public in an Age of Uncertainty* (pp. 30–49). Cambridge: Polity Press.

Ottinger, Gwen. (2010). Buckets of Resistance: Standards and the Effectiveness of Citizen Science. *Science Technology Human Values* 35 (2): 244–270.

Reusswig, Fritz. (2010). The New Climate Change Discourse: A Challenge for Environmental Sociology. In M. Gross and H. Heinrichs (Eds.), *Environmental Sociology: European Perspectives and Interdisciplinary Challenges* (pp. 39–57). Dordrecht, Heidelberg, London, and New York: Springer.

Schrecker, Ted. (1995). Environmentalism and the Politics of Invisibility. In M. D. Mehta and E. Ouellet (Eds.), *Environmental Sociology Theory and Practice* (pp. 203–217). Toronto: Captus Press.

Sellke, Piet and Renn, Ortwin. (2010). Risk, Society and Environmental Policy: Risk Governance in a Complex World. In M. Gross and H. Heinrichs (Eds.), *Environmental Sociology: European Perspectives and Interdisciplinary Challenges* (pp. 295–322). Dordrecht, Heidelberg, London, and New York: Springer.

Tindall, David B. (1995). What is Environmental Sociology? An Inquiry into the Paradigmatic Status of Environmental Sociology. In M. D. Mehta and E. Ouellet (Eds.), *Environmental Sociology Theory and Practice* (pp. 203–217). Toronto: Captus Press.

Wynne, Brian and Dressel, Kerstin. (2001). Cultures of Uncertainty-Transboundary Risks and BSE in Europe. In J. Linnerooth-Bayer, R. E. Lofstedt, and G. Sjostedt (Eds.), *Transboundary Risk Management* (pp. 121–154). London: Earthscan.

Yearley, Steven. (2010). Understanding Responses to the Environmental and Ethical Aspects of Innovative Technologies: The Case of Synthetic Biology in Europe. In M. Gross and H. Heinrichs (Eds.), *Environmental Sociology: European Perspectives and Interdisciplinary Challenges* (pp. 97–108). Dordrecht, Heidelberg, London, and New York: Springer.

Zhang, Joy Yueyue. (2015). Cosmopolitan Risk Community and China's Climate Governance. *European Journal of Social Theory* 18 (3): 327–342.

Part I

Air pollution politics in East Asia

2 Politics of air pollution

How fine dust has become a politicized issue in Korea

Minjae Kim and Dowan Ku

2.1 Introduction

Air pollution, especially the type involving fine dust, has become one of the most urgent environmental and political issues in South Korea (henceforth, Korea) over the past several years. At the end of 2013, almost every major media outlet began to fully cover this fine dust issue. Since then, fine dust has become the most salient environmental and health issue in Korea, and also one of the most important political agendas in the 2017 presidential and the 2018 local elections.

General air quality, however, has gradually improved in Korea after air quality policies were set and implemented beginning in the 1960s; furthermore, clean fuel-related and other regulations were introduced in preparation for international sports events in the late 1980s and again at the beginning of the 2000s. The first comprehensive air quality control program in the Seoul metropolitan area started in 2004. This plan significantly lowered the concentration of fine dust over the next decade.

In contrast to the lowering of air-quality indexes, public risk perception of fine dust has risen. Even though these environmental problems drew political responses from the government, countermeasures could not relieve the public discontent. According to Gallup Korea (2014, 2019), 57% of people felt "very uncomfortable with fine dust" in 2019 compared to 45% in 2014. Air pollution, specifically fine dust, has been an important political issue since the mid-2010s. How did this situation occur?

This chapter explores how fine dust became a politicized issue in Korea. We analyze the following research questions.

First, how did the fine dust issue become politicized? We review the history of air pollution and air quality policies from the 1970s to the 2010s in Korea. We also present some noteworthy events and issues related to this politicization process in recent years.

Second, what kinds of policies have been suggested to deal with fine dust and by whom? In other words, what policies have been implemented amid the contestation of social forces? In terms of pollution sources, the issue of "dust from China" versus "Korean domestic emissions" has formed the axis of the air

DOI: 10.4324/9781003211747-3

pollution controversy in Korea. Among Korean domestic sources, coal power and diesel vehicles have been the most salient issues.

Third, how has the Korean state dealt with the inherent contradictions between capitalist economic imperatives and the legitimation imperatives of public health and environmental protection? We analyze current air pollution politics by drawing on research on the relationship between the capitalist state and environmental imperatives (Jessop, 1990; Hay, 1996; Dryzek et al., 2002; Davidson, 2012).

In conclusion, we suggest the concept of "air as the commons" and analyze the politics of air pollution in terms of the limits of the developmental state.

2.2   Politicization and imperatives of the modern state

Politicization refers to the process of making something non-political into something political. The demarcation between the political and the non-political may sometimes seem to be taken for granted; however, there are no natural lines between them (Jessop, 2016: 48). Every element of society can be made political but politicizing it requires concentrated efforts or specific conjunctures. Therefore, "politicization" as a process is a kind of invention involving new perspectives and controversies (Palonen, 2003).

A working definition for politicization has, however, remained scarce thus far. The large body of work that frequently uses the term "politicization" usually does not provide a clear and working definition (De Wilde, 2011). This chapter adopts the three dimensions of politicization suggested by De Wilde (2011). The first dimension involves the politicization of the issue itself. As various social actors express differing opinions on a given issue in public debate, the issue becomes increasingly contentious and public demand for policies relevant to that particular issue grows. The second dimension involves the politicization of decision-making. The decision-making process deviates from the existing "technical" approach, and adopts a more political approach—that is, the process is subjected to more "political" bodies and comes under more pressure from various stakeholders. Finally, the institutions and organizations involved in the politicized issue are subjected to more pressure because of existing party politics and ideological cleavages.

When a new issue takes on a weighty political agenda, the state must deal with the issue. The inclusion of environmental domains under state's activities is a relatively recent phenomenon (Meadowcroft, 2012). In the developed world, it was only after the late 1960s and early 1970s that states began to make efforts to protect the environment in a more modern sense. Since then, environmental domains have grown significantly in modern states, and the complexities of environmental domains have also increased. However, different countries may adopt different paths or degrees to which they accept environmentalism and related issues. Dryzek et al. (2002) classified environmental politics based on to the different relationship between civil society and the state. State involvement in environmental conservation issues could lead to contradictions with other state activities and priorities (Meadowcroft, 2012).

According to Marxist state theory, a state faces a crisis related to legitimacy, finances, and rationality owing to contradictions between economic growth imperatives and others (Davidson, 2012). When the state faces any threat to its legitimacy as a guardian of the environment, it is likely to respond at a tactical level rather than to solve the actual contradictions (Hay, 1996: 429). While Jessop accepts this analysis, to overcome economic determinism, he presents a strategic-relational approach to observe the state as a site for contestation among various social forces (Jessop, 2002: 40, 2016: 54).

The modern state depends on economic growth, but simultaneously, capital accumulation depends on the modern state. Jessop argues that the modern state itself is a major condition for economic growth (Jessop, 2016). Because of these tight interdependencies, political programs are closely tied to economic growth imperatives (Offe, 1975, as cited in Jessop, 2016). The hegemonic bloc, which assumes territorial power at the site of the state, carries out hegemonic projects related to non-economic goals (social reform, health, etc.) to ensure economic growth and build political legitimacy (Jessop, 1990: 208). Thus, the fine dust issue, which is an environmental and public health issue, has become a threat to the hegemonic bloc in Korea. However, the state as a site has strategic selectivity with regard to particular institutions and systems, so that the state prioritizes certain interests related to the contestations among social forces (Jessop, 1990: 260).

This politicization process is discussed in the following sections. Section 2.3 traces some noteworthy events and controversies surrounding Korea's fine dust issue. Section 2.4 explores fine dust-related discourses from three leading media outlets. Drawing on these empirical observations, we then discuss the politicization process in Section 2.5.

2.3 A history of contestations and policies surrounding the fine dust issue

2.3.1 Establishment of air quality policies: 1970s–2010s

Air pollution became a serious problem due to rapid industrialization and urbanization in the 1970s and 1980s. The annual mean concentration of sulfur dioxide in Seoul in the 1980s was over 0.05 parts per million (ppm), which is 10 times higher than the present concentration level. According to a 1982 survey on public attitudes toward the environment, 30.7% of respondents perceived air pollution as a major severe environmental problem (Environmental Protection Agency of Korea, 1982).

During the 1970s–1980s, basic air quality policies, including the Pollution Prevention Act (1963) and the Air Quality Standards (1983), were established. The Environmental Protection Agency (the forerunner of the Ministry of Environment, ME) was established in 1980. The enactment of laws, such as the Pollution Prevention Act (1963) and the Environment Conservation Act (1977), helped to address environmental issues in the Korean legal system. In 1990, more detailed laws were established for dealing with issues in each medium, including clean air and water quality.[1] In 1993, the environmental standard was set for PM_{10} ($80\mu g/m^3$/year, $150\mu g/m^3$/day).

International sports events triggered the implementation of air quality management policies (Rhee and Jeong, 2003). The 1986 Seoul Asia Games and the 1988 Seoul Olympics led to the introduction of clean fuel and air quality monitoring systems and the strengthening of exhaust emission standards. Certain large commercial and public facilities in Seoul were then obliged to use liquefied natural gas (LNG) in 1986, and this requirement was later expanded to all boilers with a capacity over 2 tons in Seoul in 1988. During the Seoul Olympics, only low-sulfur diesel (with sulfur content below 1%) was allowed to be used in Seoul. The proportion of low-sulfur diesel in diesel fuel was 30–40% in the early 1980s, but it grew to 71% in 1986 and 72% in 1988. The Korean government also started cracking down on polluting cars in 1986, enacting stricter automotive exhaust emission standards.

The FIFA World Cup of 2002 also led to the promotion of some air quality schemes: adoption of the compressed natural gas (CNG) buses and ultra-low-sulfur diesel; early adoption of low-sulfur (0.3%) diesel in hosting cities; and a clampdown on cars and pollution-emitting businesses. During this World Cup, the government implemented a mandatory alternate-day driving ban[2] and a temporary pause on the operation of incinerators and power plants. Thus, major hosting cities and event periods became spatio-temporal detailed targets for air quality management during this period.

Most of these air pollution regulations targeted Seoul and its neighboring cities. Nevertheless, in the early 2000s, the Seoul metropolitan area was still the most polluted region in Korea because of its burgeoning cars, dense population, and high energy consumption. Comprehensive policies targeting fine dust began to be implemented in 2003. The Special Act on the Improvement of Air Quality in the Seoul metropolitan area was enacted in 2003 to alleviate the increasing air pollution in the Seoul metropolitan area. Under this Act, the 1st Seoul Metropolitan Air Quality Improvement Plan was implemented from 2005 to 2014. In 2007, the environmental standard for PM_{10} was strengthened to $50\mu g/m^3$/year and $100\mu g m^3$/day. In 2012, the standard for $PM_{2.5}$ was set as $25\mu g/m^3$/year and $50\mu g/m^3$/day.

The main goal of the first Seoul Metropolitan Air Quality Improvement Plan (2005–2014) was to reduce the concentrations of nitrogen oxides (NOx) (22 ppb), sulfur oxides (SOx), volatile organic compounds (VOCs), and particulate matter (PM_{10}) ($40\mu g/m^3$). To achieve this goal, the government allotted a predetermined total amount of SOx, NOx, and VOCs for Seoul, Incheon, Gyeonggi province, and some large businesses. Local governments and large businesses were required to draw up plans to reduce emissions within the allotment. Vehicular emissions are also important pollution sources; most of the air quality improvement plan budget is spent on reducing vehicular emissions. Thus, stronger car exhaust emission standards were adopted for manufacturers and existing vehicles. Highly polluting vehicles were required to attach a diesel

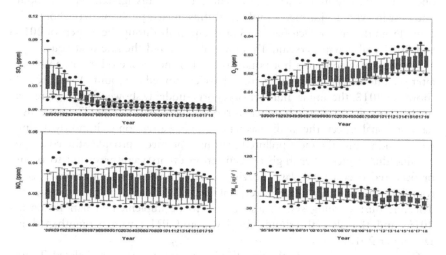

Figure 2.1 Annual mean concentration of pollutants (1989–2018)
Source: Annual report of air quality in Korea 2018 (Ministry of Environment, 2019)

particle filter or were earmarked for early scrapping. Furthermore, a subsidy was introduced to encourage the use of clean cars, such as electronic and hybrid cars. The plan cost 4 trillion Korean Won (KRW), and 80% of this amount was used for reducing vehicular emissions. This plan was partly successful in lowering PM_{10} (from 61 to 45 µg/m') but did not meet its main goal, especially with regard to NOx.

The box plot[3] in Figure 2.1 depicts the change tendency of air quality indexes. The annual ozone concentration shows a worsening trend. However, Figure 2.1 suggests that the general tendency of air quality indexes has gradually improved. The concentration of SO_2 plummeted throughout the 1990s and now satisfies established environmental standards. The concentrations of NO_2 and PM_{10} nearly met the established standards, but some observation data exceeded the established limits. Regardless of these trends, the fine dust issue has drawn immense public attention.

2.3.2   *Fine dust as an emerging issue: 2013–2016*

Although some of the planned air quality improvement goals were not satisfied, most air-quality indexes improved. Nevertheless, fine dust became visible to the public through some conjunctures, namely, high fine dust concentration events and the fine dust forecast system. The high fine dust concentration events occurred in Beijing, China in January and October 2013. Furthermore, around this time, the National Institute of Environmental Research (NIER) started to forecast fine dust concentrations. The preliminary forecast was started in August 2013 in Seoul and gradually extended nationwide in November; it started operating officially in January 2014. These two events ignited reports about fine dust in every major media outlet and thus, fueled public anxiety about this issue. From the end of October 2013 (and especially during the winters of 2013 and 2014), all newspapers and TV news fully covered the fine dust issue.

Fine dust has become a hot potato issue. It is not a one-off event, and year after year, news reports about this issue have continued to mount. As Figure 2.2 shows, in 2013, the annual number of reports suddenly showed a large increase; furthermore, in 2014, the number of reports increased by 77%. The issue also has a seasonal aspect: the media has tended to pay attention to it mostly during winter and spring, the most polluted seasons. The media provided full coverage for fine dust issues when high concentration events were predicted to occur or occurred over several days. Specific research results and controversies surrounding the fine dust issue were published in daily and online newspapers. The fine dust warning system also began to be implemented nationwide in 2015, and the alert thresholds for PM_{10} and $PM_{2.5}$ were strengthened in December 2015.

Government responses to the fine dust issue were, nevertheless, belated. There has been some bureaucratic implementation of air quality policies, but only in April 2016 did former president Park Geun-hye first respond to these controversies. The ME set up the 2nd Seoul Metropolitan Air Quality Improvement Plan (2015–2024) in 2014. This plan newly covered $PM_{2.5}$ and ozone and planned to decrease annual mean concentrations of PM_{10} to 30 µg/m³ and $PM_{2.5}$ to 20 µg/m³ by 2024. The plan's structure was similar to that of the previous plan, and stronger schemes were added to deal with issues related to non-road vehicles,[4] biomass combustion, and so on.

However, these schemes were not fully operational, and some high concentration events occurred over a few days in April 2016. On May 10, the president called for an "extraordinary countermeasure" at a cabinet meeting, and the ME

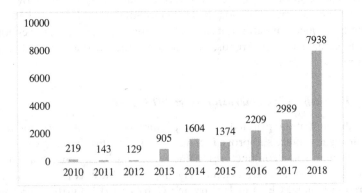

Figure 2.2 Number of articles about fine dust

Source: Bigkinds (www.bigkinds.or.kr). Keywords: Fine dust, three broadcasting systems (KBS, MBC, and SBS), and two newspapers (*The Hankyoreh* and *The Kyunghyang Shinmun*)

started to set up some countermeasures. The government announced special measures to deal with the fine dust issue on June 3 and a detailed plan on July 1, 2016. The new policy focused on reducing vehicular exhaust emissions, including those of non-road vehicles, and the early phase-out of old (more than 30 years) coal-fired power plants, among other issues; however, many critics argued that this policy was just a combination of existing policies.

After the impeachment of former president Park, the fine dust issue became a major agenda in the presidential race of May 2017. Moon Jae-in, the frontrunner and eventual winner, promised to initiate a 30% reduction in fine dust emission within his term of office by strengthening environmental standards, installing a dedicated presidential organization, and initiating a fine dust summit agenda with China. It was the most popular commitment of the elections and received the highest number of "likes" on the presidential race website.

2.3.3   *The fine dust issue as a chronic problem: 2017–present (2020)*

After the presidential election, President Moon's third work directive dealt with the fine dust issue. The first report by the ME also discussed fine dust policies, including the strengthening of fine dust management standards. The Moon administration's first fine dust control policies were announced on September 26, 2017. In line with its pledge to reduce fine dust by 30% within 5 years (the previous government's goal had been to reduce fine dust by 14% by 2021), the administration decided to abolish seven aged thermal power plants within Moon's term in the power generation sector, discussed LNG conversion for four new thermal power plants, and promised strict application of emission standards for five plants. With regard to the industrial sector, 77% of 2.86 million diesel cars are to be scrapped early in Moon's term, while the total pollution emission management system will be expanded nationwide. The environmental standards applied to fine dust levels were also strengthened to the level of advanced countries ($50 \ \mu g/m^3 -> 35 \ \mu g/m^3$).

The fine dust emergency reduction measures first went into effect at the end of December 2017, after being announced in February 2017. In March 2018, stronger standards were introduced for $PM_{2.5}$: the environmental standard was changed to $15 \mu g/m^3/year$ and $35 \mu g/m^3/day$; the $PM_{2.5}$ forecast standard denoted as "bad" was changed from 51 to $36 \mu g/m^3$, and that denoted as "very bad" was changed from 101 to $76 \mu g/m^3$. The fine dust warning system was also strengthened in April 2018. Despite such efforts, the fine dust situation did not improve, and thus, a petition was posted on the Blue House website in early 2018 to demand a diplomatic protest against China; support for this petition exceeded the government's minimum response standard of 200,000 responses on March 30. In May, the government mentioned that it would deal with the fine dust issue as a summit agenda item. In June, the China–Korea Environmental Cooperation Center was opened, and President Moon met President Xi Jinping at the 2018 Asia Pacific Economic Cooperation meetings, where President Moon mentioned the fine dust agenda.

The fine dust issue also became a major topic during Korea's June 2018 local elections. In the Seoul mayoral election, candidate Park Won-soon proposed the creation of a green transportation promotion zone that would restrict the operation of diesel vehicles that fall under the (dirtiest) Grade 5 category. Almost all the candidates promised a supply of air purifiers for educational facilities and the implementation of electric and hydrogen vehicle infrastructure, while some leading candidates pledged to install large outdoor air purifiers and 100 outdoor dust purification towers.

The government's second comprehensive plan was announced in November 2018. First, the clean diesel policy, which began in 2008, was scrapped. The policy, which was first initiated by the Lee Myung-Bak administration, promoted the use of "low-emission diesel cars" through incentives as part of its Korean Green Growth Initiative. Over 10 years, the number of diesel cars increased from 6 million to approximately 10 million in 2018. Second, the government planned to support the early scrapping of diesel vehicles by doubling of subsidies (up to 4 million KRW per car). Third, in the power generation sector, environmental costs would be reflected in power dispatch (environmental dispatch) and fuel taxes. The preliminary reduction measures are to be issued when the forecast is "very bad" or expected to exceed $50\mu g/m^3$.

In February 2019, the Special Act on the Reduction and Management of Fine Dust came into effect. Under this law, a special measures committee was established to deal with the fine dust issue, and the legal basis for introducing emergency reduction measures was also provided. Furthermore, every 5 years in the period beginning from 2020, the government is expected to establish a comprehensive plan for fine dust reduction and management, and the heads of the metropolitan and provincial governments should establish an implementation plan. The special measures committee on fine dust was formed by the prime minister Lee Nak-yon in February 2019. The committee comprises 35 members, including ministers from related ministries and experts on emissions, pollution reduction, and health.

In the winter of 2018 and the spring of 2019, fine dust levels soared. Emergency reduction measures were issued for a week between March 1 and 7, 2019. As public opinion worsened, the environment minister suggested artificial rainfall as one solution to the problem; President Moon suggested that all measures should be considered and that emergency reduction measures or artificial rainfall experiments with China should be considered. After this, extra finances were allotted, and the fine dust budget increased. In March 2019, a high concentration fine dust event was defined as a social disaster in the revision of the Framework Act on Disaster and Safety Management. As fine dust was designated as a social disaster, a fine dust crisis warning (PM2.5) was established to manage crises involving high fine dust concentrations.

In April 2019, the Special Act on the Improvement of Air Quality in the Air Management Area was enacted. This act replaced the Special Act on the Improvement of Air Quality. In addition to the existing Seoul metropolitan area, the central, southeastern, and southern populated and industrial areas were newly designated as Atmospheric Management Areas; furthermore, basic plans were established for each area, which are required to be updated every 5 years. The main goal of this plan is to reduce the total amount of pollutants by 40% by 2024 (based on emissions of NOx, SOx, and total dust).

The National Council on Climate and Air Quality (NCCA) was established on April 29, 2019; headed by former UN Secretary-General Ban Ki-moon, it was designed to act as a national body for resolving fine dust problems. Various civil society organizations, including non-governmental organizations, academic organizations, and religious organizations, participated in the committee. It gathered opinions from experts and citizens and advised the president. After discussions with the public engagement committee and experts, in September 2019, the NCCA proposed a seasonal management system for initiating and overseeing efforts to reduce fine dust emissions regularly between December and March. To implement these recommendations, the government introduced a seasonal management system in December 2019. Between December and March every year, the government will reduce the operation of public work sites, restrict the operation of vehicles falling under the Grade 5 category, expand suspensions of coal-fired power plants, and impose upper limit restrictions on coal-fired power plants (up to 80%).

The Moon administration, however, has not produced many tangible results. A poll conducted by KBS (March 2019) showed that 80% of citizens felt that the government was not doing well with regard to the fine dust issue. In the short term, it is impossible to parley with China and to reduce Korean domestic air pollutant emissions and concentrations dramatically. Thus, the fine dust issue has been and will continue to be a social and political challenge.

This chapter suggests that the fine dust issue could have been sparked by severe smog events in China and the commencement of fine dust forecasts, which were amplified by media outlets. Furthermore, belated responses from the Park administration, which were followed by mass protests because of her political scandal and impeachment as well as the early presidential elections, gave the political opportunity to raise the fine dust issue as a heavily politicized agenda. In the politicization process of the fine dust issue, the mass media played an important role to construct discourses about the issue in the public sphere of Korean society. In the next section, we explore the discourses of these media outlets.

2.4   Media discourses on the fine dust issue

The most contentious part of the politicization of the fine dust issue is determining where responsibility for the issue lies. To determine how the origins of fine dust were perceived and constructed within the public sphere of Korean society, we analyzed articles on this topic from three leading media outlets in Korea. First, we analyzed the Joongang Tongyang Broadcasting Company (JTBC) evening news, which has played an important role in the politicization of the fine dust issue. We also looked at editorials from *The Chosun Ilbo*, the most representative conservative newspaper, and *The Hankyoreh*, the most representative progressive newspaper. Our analysis data were derived from media websites, where we searched all articles and editorials by using the keyword "fine dust."[5]

The JTBC's evening news featured 607 reports on the fine dust issue from 2012 till early 2020. It is noteworthy that the JTBC's reports contained many references to China. First, China was mentioned in 320 articles (more than half of the total articles). About half of these (148 articles) emphasized the notion that fine dust came from China through the expression "from China." This emphasis on China continued throughout the 2012–2020 period.

The phrase "dust from China" began to appear during the early stages of this period. The JTBC's seven reports in 2012 and 2013 were all about smog from China. At that time, the channel reported on extreme high-concentration smog events in China, including Beijing, stressing that the fine dust flows originated in China. With the introduction of the $PM_{2.5}$ warning system (starting with Seoul in 2013), TV news shows began to broadcast fine dust forecasts in December 2013.

We divided JTBC's reports on China into four types. The first involved reports about China's local situation or China's air pollution policy. There were reports about high-concentration smog situations in China or air quality policies, including heating and factory relocation. The second type of report involved articles on scientific research about the path from and fine dust levels in China. These JTBC reports on joint research with the US's National Aeronautics and Space Administration as well as Korean researchers mainly focused on the origins of fine dust in China. The third type involved reports on Korea's diplomatic policies toward China. These mainly discussed the poor environmental cooperation between Korea and China, the Korean government's insufficient response in this regard, and the introduction of international successful cases. The last type of report involved daily weather forecast-related articles. With the implementation of a nationwide fine dust forecast system, fine dust forecasts have now become routine on the news. During this time, the phrase "dust from China" was routinely used with the following example phrase: "(fine dust levels will be 'bad' because of) the inflow of fine dust from China as well as the domestic factor along with atmospheric congestion."

Korean domestic factors became the subjects of articles mainly when government measures were taken or when environmental groups posed related questions. The reports, which dealt with various sources of fine dust, including coal-fired power generation, diesel vehicles, and manufacturing businesses, pointed out the defects in government policies. While calling for stronger policies, some of these reports also pointed out the potential negative effects of these policies on the economy, including higher diesel prices and electricity tariffs. Several reports also discussed fine dust as an important policy agenda in the 2017 presidential election and the 2018 local election.

We also analyzed editorials from *The Hankyoreh* progressive newspaper and *The Chosun Ilbo* conservative newspaper to further determine whether and how the public sphere of Korean society has constructed fine dust-related issues differently depending on ideology and political faction. *The Hankyoreh* published 51 related editorials from 1996 to 2019. *The Hankyoreh*, which used the term "fine dust from China" in 2013, called for active policymaking from the South Korean government and a responsible attitude from the Chinese government. Since then, however, the newspaper has stressed the importance of Korean domestic factors and comprehensive policies. This newspaper argued that, although China was significantly contributing toward exacerbating the fine dust issue, Korean society should focus on domestic factors and countermeasures.

In 2019, six editorials were published in *The Hankyoreh*, focusing on emissions originating in businesses, and stressing corporate responsibility and government oversight. The editorials criticized businesses that manipulated the measurement of fine dust emissions and local governments that were in charge of overseeing compliance but imposed only weak penalties. The editorials also discussed exhaust gas that originated in furnaces operating with broken fine dust prevention facilities and exhaust gas that was routinely discharged illegally. The newspaper called on the government to strengthen policies dealing with manufacturing sites, the largest sources of fine dust.

The Chosun Ilbo published 52 editorials from 2013 to 2020. It published editorials about fine dust originating in China until 2013 and 2014. In 2015 and 2016, it published a series of editorials on diesel cars and Korean domestic factors in the wake of Volkswagen's emissions manipulation scandal. This trend changed dramatically after 2017 when the Moon administration assumed office. Half of the editorials now criticized the new energy policy "nuclear power phase-out."

The Chosun Ilbo points out that, while reducing coal power generation is necessary, it is impossible to achieve a simultaneous reduction of nuclear power plants. The newspaper has insisted that the shutting down of nuclear plants would lead to increased LNG and coal power generation, which emit fine dust. After some serious fine dust concentration events occurred in March 2019, an editorial titled "Long weekend with serious fine dust, let's scrap the nuclear power phase-out policy" argued that China cannot be blamed as the sole cause of the problem and that, while all countermeasures should be taken, the government's contradictory energy policy should be changed first. In other words, it is impossible to simultaneously push for a phase-out of both nuclear and coal power; furthermore, coal-fired power generation is not decreasing because of the government's contradictory energy policy.

The analysis of the three selected media outlets yielded the following implications. First, each media outlet perceived the main cause of the fine dust issue differently. While the JTBC focused on China's responsibility, *The Hankyoreh* emphasized the importance of policies about overall Korean domestic factors, while *The Chosun Ilbo* focused on the effect of energy policies, especially nuclear power phase-out policies after 2017. Thus, we found that fine dust became a distinctive political issue after 2017. Public discontent regarding fine dust has been routinely mentioned in editorials since 2017. The editorials of *The Chosun Ilbo* and *The Hankyoreh* showed the influence of their own political positions.

Referring to the fine dust issue, *The Hankyoreh* criticized Korea's development policies for lacking environmental protection, while *The Chosun Ilbo* criticized the nuclear power phase-out policy.

2.5   Discussion

2.5.1   *Politicization of the fine dust issue*

We use three dimensions to analyze the politicization of the fine dust issue in three periods. The first period is from 2013 to 2016 when the fine dust issue first received attention from the media and civil society. During this period, the fine dust issue became a salient topic of public debate and drew strong media attention. In addition to existing environmental groups, new civic groups calling for policies to deal with fine dust issues emerged. However, air quality policies were implemented in a conventional technocratic manner. This period was marked by escalating discontent due to ineffective policies or a lack of political responses.

The second period began in May 2016, when then-president Park authorized extraordinary countermeasures to deal with fine dust; this period included the 2017 presidential election and the period after Park's impeachment. With growing public pressure for a resolution to the fine dust issue, Park responded directly. Thus, the fine dust issue emerged as an important agenda amid various political changes, including mass protests amid Park's political scandal, the impeachment, and the presidential election. From this period onward, fine dust policies were determined in a more political manner, a departure from the previous depoliticized technocratic method. Major policies were determined by a presidential directive and a pan-government task force consisting of related ministries.

The last period started at the beginning of the Moon administration and continues to the present (2020). The fine dust issue began to be affected by party politics and ideological cleavages in connection with shifts in energy policy. The Moon administration quickly formed a pan-government task force and established related legal systems and institutions. The president directly instructed the prime minister and the environment minister to deal with the fine dust issue, especially during winter and spring, when fine dust concentration levels soar. However, conservative media, including *The Chosun Ilbo*, and the conservative opposition party have tried to create ideological cleavages by linking the fine dust issue to Moon's energy policy of de-nuclearization.

2.5.2   *Characteristics of fine dust politics*

The most prominent feature of fine dust politics is that responsibilities for it are distributed and embedded throughout society. The fine dust emission inventory covers the entire daily energy use, ranging from industrial giants to automobiles and from home heating boilers to outdoor incineration. Fine dust politics embody the complexity of asking who should be held responsible for

Table 2.1 Politicization process of the fine dust issue

	Issue	Decision-making process	Party politics and ideological cleavages
1st period (2013–May 2016)	Draws attention from media outlets and civil society	Conventional, technocratic manner	
2nd period (May 2016–2017, Presidential election)	Develops as an important agenda with political upheaval	Major policies discussed in the pan-governmental task force	
3rd period (2017–present)	Continuing discontent	Directives from President/ With the task force, a special committee and national body (NCCA) was launched	Conservative social forces connect energy policy (e.g., denuclearization) to this issue

how much of the problem. Under these circumstances, China has often been cited as an external enemy since the beginning of the fine dust politicization process.

China was labeled as the main source of fine dust during the first phase of the issue. Smoggy images of China and the NIER's forecasts were broadcast continuously and helped to develop the "dust from China" narrative. This narrative is easily adopted by a vast majority of Korean people with experience of Asian dust.[6] However, unlike Asian dust, most of the fine dust has a domestic origin. National institutions and media outlets amplified the "dust from China" narrative. The NIER stated that "high concentration of fine dust is predicted because of the inflow of Particulate Matter (PM) from China and additional domestic pollution" (2013.11.12.). Furthermore, the NIER and ME repeatedly announced air pollution modeling results that showed "dust blowing from China." Almost every media outlet published numerous news articles about this. Even evening TV news weathercasters usually stated that dust from China would flow into Korea.

With a series of diplomatic disputes between Korea and China after 2016,[7] the fine dust issue became a diplomatic issue. Two-thirds of Koreans believed that overseas inflow was the dominant source of fine dust (Gallup Korea, 2019) and that cooperation with China was the most urgent measure for reducing fine dust (KBS, 2019). Only some environmental groups and experts emphasized the Korean domestic emissions and the high uncertainty of the modeling results; however, these groups faced angry denunciations from some groups of people.

Some experts and environmental groups argued that the reduction of Korean domestic emissions should be prioritized (Korean Federation for Environmental

Movement, 2018); however, they faced two difficulties. First, there are various Korean sources of fine dust emission, and it is difficult to determine the focal points of action against domestic emissions. Second, these Korean domestic sources are often deeply embedded in industrial activities and everyday practices; therefore, it was easier to arouse and amplify negative public opinion against Korean domestic measures and through the "dust from China" narrative.

In this context, policies against coal-fired power and diesel vehicles could be controversial. There has been some consensus that reducing coal power is necessary, but the pace of this reduction is still debatable. Although President Moon pledged to phase out coal power in the presidential election 2017, the installation capacity of coal power is expected to increase from 36 GW to 42 GW by 2023. Furthermore, measures, including escalating electricity prices and the introduction of environmental power dispatch, are being discussed but have not yet been implemented. Through the implementation of the seasonal management system, 15 to 28 coal-fired plants were temporarily shut down during the winter, and other coal-fired power plants also tried to reduce fine dust emissions by limiting their output to 80%.

The Park administration had planned to close 10 old coal-fired plants by 2025, while the Moon administration initially planned to close the remaining 6 old plants by 2021. However, in the eighth basic plan for electricity supply and demand (announced at the end of 2017), the proportion of coal-fired power was expected to grow from 36.8 GW in 2017 to 42 GW in 2022. Seven of the nine new coal-fired power plants will be built as planned, while two will be converted to LNG. The plan also proposed changing the current priority of electricity dispatch, which favors cheaper coal, to environmental dispatch, which reflects environmental costs; this has not yet been introduced and is being discussed in the ninth plan (2020–2034).

Debates regarding diesel vehicles are especially fierce. Stakeholders, including the nearly 10 million diesel vehicle owners and huge car businesses (13% of the manufacturing sector), have raised their voices. Mass media outlets have discussed pollutant contributions from diesel combustion cautiously compared to the "dust from China" narrative. Highly technical issues, such as the emission coefficients of wheels and engines with regard to different testing methods, have also been discussed. The energy relative price structure for LNG, diesel, and gasoline has become a most sensitive issue not only for interest groups but also among ministries. Even in the Moon administration, some independent committees[8] have recommended that the relative energy tax should be readjusted; however, the Ministry of Economy and Finance and the Ministry of Trade, Industry, and Energy (MoTIE) have opposed raising taxes on diesel because of the expected economic impact.

Certain diesel car policies have also been proposed to eliminate incentives for diesel vehicles and to raise diesel prices. The clean diesel policy, which used to give incentives to diesel cars, was abolished in 2018 under the government's fine dust measures. Measures to restrict the operation of Grade 5 (the dirtiest) diesel vehicles were also expanded. However, an increase in diesel prices was not realized. Since 2005, when diesel passenger cars were authorized for sale, and gasoline and diesel relative prices were adjusted to 100:85, energy tax system reforms have been discussed. The Park administration discussed increasing diesel prices, and four national policy research institutions jointly examined this subject. However, due to backlash from diesel fuel consumers, the Moon administration did not change diesel prices. The NCCA also proposed relative energy tax system reforms, but it designated this as a mid-term to long-term task.

Because of high public anxiety and discontent, the Korean government has had to respond to the fine dust issue; nevertheless, it has introduced a few substantial reforms for the energy and industry system. Reform policies that could disrupt economic activities received strong objections from the "development coalition"[9] (Ku, 2004). To alleviate public discontent, the government undertook some actions, such as short-term policies (e.g., seasonal management system) and the establishment of the NCCA under the former UN Secretary-General Ban. In turn, capitalist economic imperatives still overwhelm legitimation imperatives.

2.5.3 The capitalist state and environmental imperatives

Policies to deal with fine dust have not been sufficient to decrease fine dust concentrations, despite the high public interest in the fine dust issue. Most fine dust policies have focused on short-term countermeasures. As we suggested in Section 2.5.2, structural reforms in the energy and power sectors have not been sufficiently implemented.

Furthermore, Korea's policy elite, who emerged through the path of the developmental state, still have close links with industrial sectors (Kalinowski, 2021). Under an export-dependent economy, South Korean industries are sensitive to international competitiveness and, crucially, to energy prices. Furthermore, the popularity of developmental philosophy among Korea's policy elite stems from a long history of developmental state experience (Kim and Thurbon, 2015). In the pan-government task force and committee, civil society and ME-backed proposals were blocked by the MoTIE and other ministries that were closely connected with industry. Proposed increases for electricity and energy prices are repeatedly blocked, because they could burden industry or the public. Consequently, coal-fired power plants are still being built to supply cheap electricity.

While economic growth imperatives are a major social principle, other imperatives could gain predominance within a short period (Jessop, 2002: 28; Quastel, 2016). With frequent high concentration events, the fine dust issue has placed great political pressure on the hegemonic bloc in Korea. Consequently, despite concerns about the economic impact of seasonal pause of coal-fired power plants, restrictions by seasonal management systems were implemented in earnest. Some areas, such as Chungnam province, were more politically successful than others— it shut down coal-fired power plants early and blocked the construction of new ones in 2018. In this sense, social movements for air quality are particularly impactful in air pollution politics.

However, in terms of long-term planning, which is determined by pan-governmental committees or energy and power sector expert committees, the existing policy elite connected to industry has remained powerful. To impose environmental imperatives on the state, counter-hegemonic projects (Quastel, 2016) are necessary; however, it seems that such alternatives are relatively weak.

2.6   Conclusion

Fresh air is universally necessary for sustaining life and survival. Therefore, air can be considered as the commons for everyone. However, as it is difficult to instate boundaries and exclude individuals from using air, it is recognized as a free good. When severe air pollution invades air as commons, such polluted air threatens the lives of many weak individuals. Air privatization has resulted in health damage worldwide. To preserve clean air, commoners or managers must manage air as commons and solve the problem of free-riders. Governments usually assume the responsibility for managing clean air. Concurrently, they must solve issues of capital accumulation or economic growth. In this context, a civil society with cautious citizens who are aware of air pollution could help to preserve clean air.

In Korea, air pollution has been a social and environmental problem since the 1970s. The government and business sectors were key air polluters during the period of industrialization in the 1960s and 1970s. As part of preparations for international sports events in the 1980s and early 2000s, however, the government attempted to control air pollution. As public concerns about air quality increased in the mid-2010s, the government was pressured to relieve air pollution.

Air pollution, specifically the fine dust issue, transformed into a political issue in the mid-2010s. Fine dust was constructed as a symbol of health hazards. Although air quality was very poor in the 1980s and 1990s, the environmental movement did not strongly focus on the air pollution issue. In the 2010s, citizens who were very concerned about "fine dust" emerged in civil society. While air pollution politics was driven by bureaucrats and technocrats before the 2010s, it was activated by politicians and citizens beginning in the 2010s.

First, we found that the fine dust issue began to be politicized when air pollution in China and the NIER's forecasting were hot issues in 2013. The media has reported many articles since 2013. Eventually, the issue became one of the weightiest agendas in the presidential election in 2017. Air as commons was managed by governments from the 1960s to the 1990s, and pollution became a citizens' issue in the 2010s. The Korean government had to improve its policies in response to people's grievances and concerns.

Second, the mass media labeled China as a major source of fine dust; by contrast, Korean domestic sources and institutions, including big businesses' emissions and environmentally harmful subsidies, did not receive much social discussion. The "development coalition" mainly criticized China and the "nuclear phase-out policy," not the Korean domestic industrial sectors. By contrast, environmental civil groups urged the government to make the industrial structures that caused air pollution and climate change more sustainable.

Third, we analyzed the state and civil society characteristics with regard to air pollution. The Korean state as a capitalist and developmental state successfully achieved rapid economic growth after the 1960s. Facing air pollution, Korea responded passively in the 1970s and early 1980s. However, it made its position more proactive in the late 1980s to successfully organize international sports events. Because civil society has been cautious about the fine dust issue that emerged in the 2010s, the state should prioritize environmental imperatives to preserve its legitimacy. However, its policy innovation has been limited, because it is based on an export-oriented fossil fuel-based capitalist economy.

We must understand that Chinese and Korean people share air as commons. Many air pollutants flow from China to Korea, and some pollutants flow from Korea to China. However, it is difficult to establish the concept of a "risk community" against the sharing of common polluted air in East Asia. Few commoners have tried to cooperate to achieve clean air across East Asia. People's thinking in this regard is often embedded in nation-based economic growth imperatives. China has been socially constructed as a polluter—not as a collaborator—by the developmental coalition. Thus, we have a long way to go before we can build an environmentally sound East Asia.

Notes

1 Later, the Natural Environment Conservation Act (1991) and the Soil Environment Conservation Act (1995) were enacted.
2 The government's mandatory alternate-day driving ban prohibited driving every other day; it could lower road traffic in half.
3 The box plot shows the maximum, minimum, upper quartile, lower quartile, median, and outliers. With the range of observations in these plots, we can observe the trend of air quality indexes more accurately.
4 The term "non-road vehicles" covers a very wide range of machinery that is usually not used on public roadways. It consists of construction machinery (loaders, bulldozers, etc.), agricultural and farming machinery, and small gardening equipment.
5 We analyze all searched articles of JTBC and editorials of *The Hankyoreh* and *The Chosun Ilbo*. The JTBC, which was founded on December 1, 2011, made reports on fine dust from 2012. *The Hankyoreh* editorials cover the "fine dust issue" from 1996. *The Chosun Ilbo* editorials started to deal with this issue from 2013."
6 Asian dust (also known as Yellow dust or China dust storms), a meteorological phenomenon that brings the dust of deserts of China and Mongolia to Korea, Japan, and the Russian Far East, is a well-known seasonal event in Korea. In this respect, the "dust from China" narrative is not a novel notion.
7 THAAD ((the Terminal High Altitude Area Defense) was the most sensitive issue on a relationship between the Korean and Chinese government. In 2016, the Korean government announced the deployment of THAAD system and it caused the reprisal of the Chinese government: "Korea limitation order" (Chinese: 限韩令). This China's so-called "wolf warrior" approach to diplomacy ignited the anti-China sentiment in Korea. With this issue, there have been a bunch of anti-China issues. According to a 2021 survey of Korean weekly newsmagazine, *Sisain*,

75.9% of Korean feel negative with China (Sisain, 2021). A 2020 survey of East Asia Institute also show that 40.1% of Korean have negative feelings to China; in 2015, it was only 16.1% (East Asia Institute, 2020).

8 These independent committees were the special presidential committee of public finance reform and the NCCA.

9 The developmental coalition is a coalition of social forces that designs and leads the development of Korea. The coalition is organized mainly around economic interests, but it may also use ideological justification to mobilize resources (Ku, 2004).

References

Davidson, Stewart. 2012. The insuperable imperative: A critique of the ecologically modernizing state. *Capitalism Nature Socialism* 23(2):31–50

De Wilde, Pieter. 2011. No polity for old politics? A framework for analyzing the politicization of European integration. *Journal of European Integration* 33(5):559–575

Dryzek, John S., Christian Hunold, David Schlosberg, David Downes, and Hans-Kristian Hernes. 2002. Environmental transformation of the state: The USA, Norway, Germany and the UK. *Political Studies* 50(4):659–682

East Asia Institute. 2020. *Global identity and Korea's diplomacy.* www.eai.or.kr/main/program_view.asp?intSeq=20030&code=70&gubun=program

Environmental Protection Agency of Korea. 1982. The report on public attitudes towards the environment Fouilleux, Eves, Jacques de Maillard, and Andy Smith. 2005. Technical or political? The working groups of the EU Council of Ministers. *Journal of European Public Policy* 12(4):609–623

Gallup Korea. 2014. *Public attitudes on fine dust.* www.gallup.co.kr/gallupdb/reportContent.asp?seqNo=532

Gallup Korea. 2019. *Daily opinion No. 339* (4th week, January 2019). www.gallup.co.kr/gallupdb/reportContent.asp?seqNo=982

Hay, Colin. 1996. From crisis to catastrophe? The ecological pathologies of the liberal-democratic state form. *Innovation: The European Journal of Social Science Research* 9(4):421–434

Jessop, Bob. 1990. *State theory: Putting the capitalist state in its place.* Polity Press: Cambridge

Jessop, Bob. 2002. *The future of the capitalist state.* Polity Press: Cambridge

Jessop, Bob. 2016. *The state: Past, present, future.* Polity Press: Cambridge

Kalinowski, Thomas. 2021. The politics of climate change in a neo-developmental state: The case of South Korea. *International Political Science Review* 42(1):48–63.

KBS. 2019. *KBS public poll.* https://news.kbs.co.kr/news/view.do?ncd=4166047

Kim, Sung-Young and Elizabeth Thurbon. 2015. Developmental environmentalism explaining South Korea's ambitious pursuit of green growth. *Politics & Society* 43(2):213–240

Korea Ministry of the Environment. 2019. *Annual report of air quality in Korea 2018.* https://www.airkorea.or.kr/web/detailViewDown?pMENU_NO=125

Korean Federation for Environmental Movement. 2018. *China contribution to high fine dust level events is lower than ME argued.* http://kfem.or.kr/?p=190057

Ku, Do-Wan. 2004. Development coalition and green solidarity: A discursive analysis. *ECO* 7:43–77

Meadowcroft, J. 2012. Greening the state? In P. Steinberg and S. van Deveer (Eds.), *Comparative environmental politics: Theory, practice, and prospects*: 63–88. The MIT Press: Cambridge, MA; London, England

Offe, Claus. 1975. The theory of the capitalist state and the problem of policy formation. In Leon N. Lindberg, Robert Alford, Colin Crouch, and Claus Offe (Eds.), *Stress and contradiction in modern capitalism*: 124–144. D.C. Heath: Lexington Books, KT

Palonen, Kari. 2003. Four times of politics: Policy, polity, politicking, and politicization. *Alternatives: Global, Local, Political* 28(2):171–186

Quastel, Noah. 2016. Ecological political economy: Towards a strategic-relational approach. *Review of Political Economy* 28(3):336–353

Rhee, Jeong-Jeon and Hoi-Seong Jeong. 2003. Dynamics of environmental policy development in Korea: How did the policy windows have been opened? *Journal of Environmental Policy* 2(1):1–30

Sisain. 2021. *Who is the core group that hate everything of China?* (21.6.17). www.sisain.co.kr/news/articleView.html?idxno=44821

3 The sovereignty of air pollution?

The political ecology of particulate matter

*Sanghun Lee**

3.1 Purpose of the research

The purpose of this chapter is to explain the politicization of particulate matter (including atmospheric particulate matter with the diameter of 10 or 2.5 micrometers, otherwise known as "PM10" and "PM2.5," hereafter referred to collectively as "PM"), which has emerged as a national concern in South Korea, from the perspective of political ecology. In other words, this chapter will examine how PM has transformed from a "natural" element into a political issue, and how the political landscape surrounding PM has been reconstructed through different strategies. PM has certainly existed in the past, but it has not been the subject of special policy considerations, and statistics about PM2.5 have only been available since 2015. However, as PM has rapidly emerged as a social concern, it has become politicized. In other words, it has become a controversial issue as well as a point on political agendas.

This chapter will explain the process of the politicization of PM from the perspective of political ecology. For this purpose, it adopts the *state-nature* relationship and methodological cosmopolitanism as theoretical concepts. The "political ecology of state-nature" (Whitehead et al. 2007) refers to the spatial strategies used in the process of politicizing PM by various actors within and through the nation-state. However, since the controversy surrounding PM has grown beyond the territorial scope of the nation-state, a broader explanation is necessary. For this purpose, Ulrich Beck's concept and perspective of "methodological cosmopolitanism" will be used (Beck and Sznaider 2006; Beck 2009; Beck and Grande 2010). Methodological cosmopolitanism offers an alternative perspective against the bounded "framing strategies" of the nation-state.

3.2 Theoretical resources

The field of political ecology began with an interest in explaining changes in natural phenomena—such as soil erosion—as political and economic changes occurring at various scales. Although there are various branches of theories

* This chapter is based on my paper that was published at *ECO: The Korean Journal of Environmental Sociology* 25(2): 7–46 (2021)

DOI: 10.4324/9781003211747-4

within political ecology, political ecologists all maintain the tradition of critiquing the paradigm of modern social science, which presupposes a dichotomy between nature and society, instead of combining nature and society in an integrated way, as "nature-society" (Blaikie 1985; Blaikie and Brookfield 1987; Bryant 2015; Swyngedouw 2015).

One of the prominent theories that aims to overcome the dichotomies of "natural science-social science," "subject and object," "science and art," or "human and non-human" is Bruno Latour's actor-network theory (ANT) (Latour 1997). Latour argues that these modern dichotomies do not see the interdependence and interrelations between nature and society, subject and object, the material and the spiritual. On the other hand, theories that appear to be heterogeneous become hybrids or assemblages by technical mediation or translation. However, ANT has been criticized for eliminating the structural differences of the "state" from other actors in society. In other words, ANT treats the state as simply another actor.

In fact, however, states have hierarchical characteristics that make them distinct from other actors. Mark Whitehead and his colleagues have tried to explain "the state" in a different manner while still using ANT. They have proposed the concepts of "state-nature" and "human-non-human relations," which derive from ANT. In their discussion of Latour, Whitehead et al. point out that the state can be viewed as a "socio-ecological network" or as a "framing device." In addition, utilizing Bob Jessop's theory of the state as a strategic network of relations, Whitehead et al. explain that states perform two different functions: producing capital accumulation and hegemonic projects. In the course of performing these functions, Whitehead et al. claim that states bring nature into their strategic relational networks and adopt particular framing strategies (Whitehead et al. 2007: 52–53). "Framing is a process that involves the 'bracketing off' of the things and objects interacting in a certain context" (Callon 1998: 249; Whitehead et al. 2007: 14 recited). A state's "framing strategy" provides a specific perspective on nature and a particular frame of perception.

I argue that there are two types of spatial strategies within a state's framing strategy: "centralization" and "territorialization." Centralization is a strategy that attributes responsibility for managing nature to the central or national government. The territorialization strategy is based on a spatial separation scheme of using nature for controlling and regulating it. Through these framing processes, nature is abstracted. In other words, nature is detached from its ecological, social, cultural, and local contexts and mobilized as a set of resources for the central state. The term, "state-nature," is used to express nature abstracted in this way by the state.

> We define state-nature as those strange but inherently familiar fragments of nature that have been politically removed and abstracted from their ecological contexts. Understood on these terms, fragments of state-nature include: the various ecological emblems that are routinely used to represent national political communities (including plants, animals, and scenery); particular

places/landscapes of ecological significance (such as national parks and nature reserves); ecological phrases, narratives, and myths (incorporated into foundation legends and nation-forming stories); territorial maps and land use surveys; and even micro-biological organisms and molecules whose transformation is regulated by various national laws and restrictions.

(Whitehead et al. 2007: 2)

This definition of "state-nature" by Whitehead et al. fits the case of PM in South Korea quite well. However, Whitehead's "state-nature political ecology" is not free from the traps of methodological nationalism. In fact, the PM controversy in South Korea extends beyond the country's national boundaries, extending in particular to China. Therefore, I propose integrating the political ecology of state-nature with Ulrich Beck's concept of "methodological cosmopolitanism" (Beck and Sznaider 2006; Beck 2009; Beck and Grande 2010). Ulrich Beck converted Kant's "philosophical cosmopolitanism" into a sociological concept and uses it as a theoretical tool to explain a global risk society in the twenty-first century. In other words, the traditionally defined "society" that was equated with the nation-state was named, "methodological nationalism," and in the context of the globalization of economies, information, violence, and risk, it is no longer considered a valid frame for explanation. For individuals living in the face of global economic risks, terror risks, and climate change risks, cosmopolitanism has indeed become a reality (Jung 2012: 201). Ulrich Beck coined the concept of methodological cosmopolitanism to refer to viewing the entire world as a risk community (Beck 2006). Another important aspect of cosmopolitanism is that it requires a new, global sphere of politics against the global economy, as contemporary economic issues play a more powerful role in politics than the nation-state, which has lost much of its political functions when faced with global risk issues. This new sphere of politics has been redefined rather than extinguished. Beck presented an analytical framework that explains the "reality dimension" (empirical and analytical theories) and the "value dimension" (normative and utopian political theories) based on both the national and cosmopolitan perspectives (Beck 2006; Beck and Sznaider 2006; Beck 2009; Beck 2011).

In order to explain the process of politicizing PM in South Korea and to suggest effective policies, it is necessary to overcome methodological nationalism and normative nationalism, which only match the boundaries of the sovereign nation-state and environmental pollution at the national scale, respectively. At minimum, based on cosmopolitan realism, it is necessary to point out that

Table 3.1 Ulrich Beck's analytical frameworks for a global risk society

	Nation-state perspective	*Cosmopolitan perspective*
Reality dimension	Methodological nationalism	Cosmopolitan realism
Value dimension	Normative nationalism	Reflexive cosmopolitanism

Source: Ulrich Beck (2011)

pollutants such as PM are not generated only in specific countries, but have a causal relationship that transcends national boundaries and to explore ways to solve PM as a transborder issue. Furthermore, in order to find a fundamental solution to the PM problem, there must be a complete reorganization of the modern industrial structure as well as a transformation of the socio-ecological system based on the perspective of reflexive cosmopolitanism. In other words, drawing from methodological cosmopolitanism, it is necessary to dismantle the nation-state's framing strategy and show that PM is a cosmopolitan reality.

3.3   The status of PM and related policies in South Korea

PM comes from many sources, such as the emissions of cars, ships, and machinery; the combustion of oil or coal; the incineration of waste and tire strips; and construction processes. Such "primary generation" of PM leads to secondary chemical reactions or the "secondary generation" of PM. In fact, "secondary-generation" PM accounts for 70% of the total generation of PM (Cheon et al. 2021: 53). Unlike general dust, PM is absorbed into people's bodies and causes various diseases, including inflammation. The International Cancer Research Institute (IARC), under the World Health Organization (WHO), designated PM as a group 1 carcinogen in 2013. PM exacerbates or causes cardiovascular disease (such as ischemic heart disease, arrhythmia, heart failure, arteriosclerosis, etc.), cerebrovascular diseases (such as strokes), respiratory diseases (such as chronic obstructive pulmonary disease, asthma, etc.), and diabetes. It is known to be involved in and to contribute further to the progression of various inflammatory reactions in the human body.

The Ministry of Environment of South Korea has released data on outdoor air quality nationwide (www.airkorea.or.kr). According to the statistics, PM10 in Korea has been gradually decreasing since first being measured in 1995, and the pollution level in 2019 was the same as during the previous year, $41\mu g/m^3$. In 2019, PM2.5 also recorded $23\mu g/m^3$, the same as during the previous year.

The South Korean government's comprehensive measures for PM management are being implemented in two categories: (1) short-term measures (2017–2018) and (2) mid-term and long-term measures (2018–2022). The short-term countermeasures include "priority implementation of emergency reduction measures in response to high concentrations of PM," and the "implementation of meticulous protection measures, with the highest priority being to protect the health of vulnerable groups (elderly people, children, pregnant women, etc.)." Mid and long-term measures include policies such as: the "dramatic reduction of PM across all sectors of society" and "preparing effective measures to reduce the impact of PM from abroad through enforcing international cooperation." On February 15, 2019, the government enacted the "Special Act on PM Reduction and Management," which includes the following measures: ① the strengthening of countermeasures for cities with high concentrations of PM (e.g., closing schools, shortening class hours); ② specifying vulnerable groups

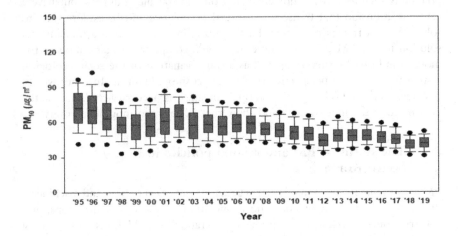

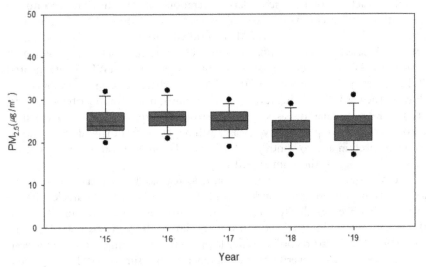

Figure 3.1 The annual average concentrations of PM10 and PM2.5 in South Korea
Source: Korea Institute of Environment and Science (2020: 19)

and areas to PM (e.g., installing air purifiers in crowded areas); ③ instituting
penalties for not implementing emergency reduction measures (if not implemented, fines of 100,000 to 2 million won are imposed); ④ the government
establishing a comprehensive plan for PM management every 5 years (with each
city/province establishing an annual implementation plan) and reviewing the
plan and its performance in cooperation with local government offices; ⑤
the strengthening of a dedicated organization for PM; ⑥ the establishment of
the "National Climate Environment Commission for Resolving the PM Problem," under the direct control of South Korea's president (April 29, 2019)
(Ministry of Environment 2019a).

3.4   The politicization of PM

3.4.1   *The social perception of PM*

In fact, despite the overall decrease in the concentration of PM pollution, according to a survey by the Ministry of Environment, public perception of the overall air quality is generally negative. First of all, compared to five years ago, only 9.7% of the general public and 11.7% of experts perceived that the air conditions in Korea were "better." In particular, 39.7% of experts recognized that the air conditions were "worse" compared to those of 5 years ago. Public concerns about PM seem to have increased significantly from 2013. There are several reasons. First, starting in 2013, PM2.5 was designated as an air quality reference material, and a forecast system was implemented. Second, the number of reports of PM in the media has exploded since 2013. Third, the WHO also announced that PM is a group 1 carcinogen in 2013 (Yeo and Kim 2019: 249). As these factors overlap, it is true that public concerns about PM have grown in general since 2013. Moreover, in March 2019, PM was declared a "social disaster" by the government. In that case, national budgets, such as reserve funds, could be used to deal with PM. Therefore, PM has been recognized as a serious social problem.

The responses of civil society groups have also been noteworthy. For example, a voluntary civic organization named, "We Ask for PM Countermeasures" (hereafter known as WAPC), was established to respond to PM as a major issue. It has 98,112 members as of July 2021 and their main activities are: sharing information about PM, suggesting policies, making campaigns, and organizing mass rallies. According to the group, several policies were enacted as a result of their activism.[1] Public anxiety and concerns about environmental hazards are called "risk perception," although risk perception does not necessarily coincide with the actual risks. In general, risk perception remains at a low level compared to the actual risk level and then increases rapidly due to specific events, but most of them converge at the actual level of risk. However, if the government or society fails to respond adequately to certain environmental hazards or does not clearly explain the cause and nature of the hazards, irrational fears may persist for a long time (Jang 2019: 209–210). However, regardless of the fact that the pollution concentration of PM has actually decreased, the reason PM has become an important point on political agendas should be explained. I would like to explain it through the perspective of the framing strategies of the state.

3.4.2   *The centralization strategy*

PM policy should be a sub-level category of air pollution management policy. However, it has been treated as the highest-priority environmental policy since 2013, without providing clear evidence that PM is more dangerous than other air pollutants. The production of standard knowledge about PM (including its quantities and estimation of emissions, etc.) is monopolized by the nation-state, giving the impression that "air is controlled by the state." However, there are many uncertainties about PM. For example, 70% of emissions come from "secondary-generation" PM, and data from the individual locations of industries within metropolitan areas is incorrect. On top of that, various variables such as temperature, wind speed, wind direction, atmospheric condition, friction power on land and ocean, make it difficult to know the exact amount of PM discharged. In other words, PM has "emergent characteristics." Because of its complexity and emerging nature, it is reasonable to regard PM data as only probabilistically accurate. Despite such uncertainties, knowledge and information about PM from the government is represented with scientific certainty. Borrowing the terminology of James C. Scott, the nation-state has increased PM's "legibility." In addition, the South Korean government has continued to argue that China has a large influence on the production of PM pollution. The Ministry of Environment and the National Academy of Environmental Sciences have suggested that China's contribution to pollution reaches 30–50% at normal times and 60–80% at high concentration times. However, controversy has continued over the reliability of the figures (Song-Chae 2019). From May to June of 2019, there was a joint study conducted by the Korea National Environment Institute and America's National Aeronautics and Space Administration (NASA) to investigate the air quality of South Korea (KORUS-AQ). It was the largest scientific study on PM, and 580 atmospheric scientists participated in the study. Table 3.2 shows part of the results of the KORUS-AQ study and shows that there are significant differences in regional levels of PM depending on the atmospheric conditions. In other words, when the atmosphere is dynamic or blocked, the contribution of PM from Korea appears to be higher, and when the atmosphere is stagnant or polluted, the contribution of PM from China (various regions) is relatively higher.

Thus, regional contributions of PM may vary depending on the meteorological conditions or various variables. Regarding the contributions of PM to other countries, the most credible research result is the "The Report of the International Joint Research Project for Long-range Transboundary Air (LTP) in Northeast Asia," published on November 29, 2019. According to the report, as a result of analyzing the domestic and international impact, the self-contribution rate

Table 3.2 Regional contributions of PM2.5

	Atmospheric Condition			
Regional Contribution (%)	Dynamic	Stagnant	Polluted	Blocked
Korea	50	34	26	57
Beijing	9	16	10	6
Shandong	17	39	38	17
Shanghai	6	2	18	0

Source: Choi et al. (2019)

was found to be 51% in Korea, 91% in China, and 55% in Japan on an annual average basis (Ministry of Environment 2019b). These findings show that there is uncertainty in determining whether PM pollution comes from inside or outside of the country. Therefore, there is a need to encourage basic pollution source management efforts, transformation of the industrial structure, and fossil fuel use reduction.

The heavy reliance of mass media on information sources from central government and experts also shows that the mass media has also been involved in this centralization strategy. According to research on the framing of mass media concerning PM, 55.4% of the information has come from government-related institutes and experts (Kim et al. 2015). It means that the perspective of mass media is biased and that the opinions of the central government (usually, a conservative one) have been overemphasized.

3.4.3 The territorialization strategy

PM has no nationality in itself. However, in the process of becoming "state-nature," it has come to have a nationality. The central government omits the scientific complexity and uncertainty surrounding the generation of PM (involving the weather conditions, the difficulty of collecting data from various sources on sea and land, and the possibility of there being various different scenarios due to assumptions in modeling). Emphasizing the impacts from China, there arises the possibility of avoiding or abandoning responsibility by the central government attempting to territorialize the air. Through this process, the problem of PM has actually become a political and diplomatic one, stimulated nationalist sentiments, and produced a tense confrontational atmosphere between South Korea and China. In 2020, active civic movement groups, such as the WAPC, announced that the main reason for the delayed solution to the PM problem was its meteorological effects and the impact originating from other countries.[2] A lot of data and graphics of mass media emphasized clearly that high concentrations of PM originated in China as well as had a profound effect on raising citizens' awareness of PM as harmful(Hwang 2018). What is even more worrisome is that if PM's scientific complexity is neglected and the Chinese factor is over-emphasized, unnecessary fears might be created. In particular, the mass media announced that PM from China would "attack" the Korean peninsula. It even used sensational terms such as: "emergency," "horror," "danger," and "fatal." Reports on the dangers of PM from China were enough to instill excessive fear of PM in the public (Jang 2019: 111–112). Eventually, a civic group filed a lawsuit demanding compensation for the damages caused by PM to the Chinese and Korean governments. The seven plaintiffs, including the leader of the Environment Foundation, claimed about China that, "As a member of the international community, it has not managed pollutants (PM) within an acceptable range, which is a violation of international norms." Also, regarding the Korean government, the leader argued that "the government has not even been able to determine accurately what is the cause of the PM," and that, "the government has neglected its duty to protect the people's right to seek safety and happiness" (Choi 2017).

While the Korean mass media over-emphasized the risks of so-called "PM from China," and the central government promoted the installation of air purifiers, the implementation of a system of driving restricted to odd-even days, and experimented with artificial rainfall, rather than providing a comprehensive and scientific explanation about PM, the public developed anti-China hate sentiments and the belief that PM is the most serious and dangerous risk. This situation can be said to be the main reason for the rapid politicization of PM since 2013. In other words, due to the strategies of centralization practiced by the national government and the South Korean media (both intended and unintended), PM, as an element previously known to be "natural," has rapidly transformed into a political and diplomatic issue.

3.4.4   Evaluation of the current PM framing strategy

This framing strategy (involving the strategies of centralization and territorialization) mobilized within and through the state, by various actors, such as the central government, the South Korean mass media, and some domestic civic groups, has narrowed down the channels for cooperation with other countries. It has also resulted in the transformation of scientific issues into political and diplomatic issues. Most importantly, it has made us forget that PM is a cosmopolitan reality. If we claim that 30–80% of the causes of PM pollution are attributed to China, and then ask to conduct joint research on PM with China, there is no way that China will respond to such a proposal. Naturally, China has insisted that South Korea's PM problem should be solved in Korea first, and in January 2019, the Ministry of Ecology and Environment in China insisted that Korea should seek its own solution for PM. Unilateral announcements on the behalf of the Korean government inevitably act as obstacles to environmental diplomacy.

What is even more problematic is that the Korean government's PM measures are distorted. Currently, the government of South Korea has provided countermeasures for high-concentration times and areas, policies catering to increased public concerns, and policies with an emphasis on human health. Such policies could distract people from focusing on a fundamental solution to the PM issue. In fact, it is difficult to achieve the environmental standards pledged by the government through such ephemeral measures. For a fundamental solution to the PM problem, measures to reduce the operation of diesel vehicles, overall changes in the fossil fuel-dependent power generation industry, reorganizing the city to increase pedestrian and bicycle transportation over automobile traffic, and managing the emissions of industrial complexes and individual sites should be implemented strongly and continuously. However, such measures have not been prepared yet. Instead, individual citizens have been responding to the problem of PM by purchasing air purifiers, resulting in the commercialization of the air by capital, rather than a fundamental transformation of the industrial structure.

Therefore, critical intervention in this framing strategy is needed. For such critical intervention to occur, it is necessary to look at the problem of PM from the perspective of "methodological cosmopolitanism." In other words, it is important to recognize that it is useless to give sovereignty to the air in analyzing the generation and movement of PM. Since there are various variables, we must acknowledge some degree of scientific uncertainty. At a fundamental level, we need to discuss how to transition our fossil fuel-dependent industrialization and urbanization processes. In civil society groups, rather than filing lawsuits from nationalist standpoints, due to having been swayed by a particular nation-state-centered framing strategy, local civil society actors should claim that the issue of PM is part of the global risk system of fossil fuel-based energy production and consumption and that our lifestyles must be transformed. In other words, a *cosmopolitan framing strategy* should be developed.

3.4.5 Suggestions for a framing strategy based on methodological cosmopolitanism

In order to become a counter-frame against an existing one, one has to reduce one's dependency upon the central government's and experts' efforts to increase the legibility of PM. Building a counter-frame will require viewing the PM issue not just as a problem of the national scale, but rather, as a cosmopolitan reality. In particular, people across East Asian countries suffer from PM together and residents of the entire region should work together to overcome fossil fuel-dependent industrialization and vehicle-oriented urbanization. Such a region-wide effort entails transborder solidarity that extends beyond the territories of nation-states in order to tackle the PM issue throughout the region. As PM is a cosmopolitan reality, we have to encourage *trans-border solidarity* as well as construct a *breathing community* out of the entire East Asian region in order to construct a fundamental solution for PM, rather than simply continuing to adhere to a framing strategy in which PM is made into state-nature at the national scale.

However, when trying to construct a breathing community in the East Asian region, we have to pay attention to the fact that international air pollution cooperation among nation-states limits our abilities to practice trans-border solidarity. As international cooperation is based on the strategy of making state-nature at the national scale, there is always the possibility of conflicts of national interests. In particular, ongoing historical scars and conflicts (i.e., due to the legacies of imperialism, war, and national divisions) among the countries of East Asia have prevented them from achieving solidarity. Further, at the same time, practicing transborder solidarity among civil society groups in East Asian countries seems inappropriate. Currently, China is an authoritarian regime that prohibits criticism about its environmental policy internally from civil society groups and the mass media (Park 2021).

Considering such conditions, an "inter-city PM policy coalition of East Asia" might be a more appropriate measure for constructing a methodological cosmopolitan framing strategy for the PM problem. Cities themselves are relatively free from ideological confrontations and conflicts of national interest. For example, in recent years, even when there emerged a serious conflict between the national governments of China and South Korea, owing to the installation of the THAAD (Terminal High Altitude Area Defense) shield in Korea, cooperation between Beijing and Seoul continued stably (Hwang and Paik 2020). In fact, in 2016, the East Asia Clean Air Cities (EACAC) initiative was established as a regional partnership platform for exchanging knowledge and policies related to air pollution control among local governments. It included ten cities in four countries: China, Japan, Mongolia, and South Korea. In the case of Seoul and Beijing, an environmental team was formed in 2014 by the mayors of Beijing and Seoul with a specific focus on air pollution, and which committed to joint research on air quality management (Seoul Metropolitan Government 2021). Similar to the case of Beijing and Seoul, other cities in East Asia can construct transborder coalitions for resolving the PM problem and for enabling concrete cooperation. Through such processes, an alternative framework that transcends the state-nature framework at the national scale might then come forth. In other words, nationalistic frameworks for the air as the problems of South Korea or China only might then be transformed into a cosmopolitan framework for the air of the entire East Asian region, constructing an East Asian breathing community in order to begin the process of practicing transborder solidarity.

3.5   Conclusion

PM has become a significant national concern compared to other air pollutants over the past decade. The South Korean government has also given priority to PM-focused policies, as compared to policies that regulate daily air pollution in general. The central government has mainly been clinging to short-term measures for high-concentration times and to mitigate heavily polluted areas. In addition, policies focusing on individuals, such as the "odd-even license plate digits policy" have been implemented in an attempt to control the number of cars on the roads each day; however, this policy merely caused inconveniences to individuals rather than had any significant, long-term effects to PM levels. Individuals also mainly depended on their own efforts (wearing masks, purchasing air purifiers, setting up air purifying systems in their homes, etc.), and these efforts resulted merely in the tendency to commercialize the air. Despite the high degree of uncertainty as to whether 70% of PM is indeed secondary-generation, the unnecessary emphasis on external factors producing inflows of PM (in particular from China) has produced nationalistic reactions during high-concentration times.

In this chapter, I have argued that PM was politicized by particular framing strategies: the centralization strategy and the territorialization strategy. As a result of such politicization, PM has become a form of state-nature at the national scale, and this framing has prevented a search for fundamental solutions to the PM problem. In other words, central government efforts to increase the legibility of PM by monopolizing scientific knowledge and removing uncertainty have caused political and diplomatic problems in the East Asia region as a whole.

In order to create a fundamental, structural solution for the PM problem, a framing strategy based on methodological cosmopolitanism is needed as a counter-framing strategy. Recognizing that PM is not a problem that can be attributed only to individual countries—but is in fact a cosmopolitan reality for all of East Asia—is a framing strategy that can put the focus on building an East Asian breathing community and a transborder solidarity practice that goes beyond nation-state units. Rather than attempting the nearly impossible task of air environment cooperation between nation-state governments or within international organizations, it is more feasible to achieve air environment cooperation between cities. Indeed, the city is an important unit and cause of the PM problem. Therefore, I argue that the counter-framing strategy of forming an East Asian urban breathing community will play a crucial role in mitigating and solving the PM problem in East Asia in the future.

Notes

1 For more information about the organization, please refer to https://cafe.naver.com/dustout.
2 https://cafe.naver.com/dustout (Announcement about the 5th anniversary of the WAPC).

References

Air Korea (www.airkorea.or.kr).
Beck, U. 2006. *Cosmopolitan Vision*. Cambridge: Polity Press.
Beck, U. 2009. *World at Risk*. Cambridge: Polity Press.
Beck, U. and E. Grande. 2010. "Varieties of Second Modernity: The Cosmopolitan Turn in Social and Political Theory and Research." *The British Journal of Sociology* 61 (3): 409–443.
Beck, U. and Chansook Hong (trans). 2011. *Segehwa Sidaeeu Kwonryukgwa Daehang Kwonryuk (Macht und Gegenmacht im globalen Zeitalter)*. Seoul: Gill Press.
Beck, U. and N. Sznaider. 2006. "Unpacking Cosmopolitanism for the Social Sciences: A Research Agenda." *The British Journal of Sociology* 57 (1): 1–23.
Blaikie, P. 1985. *The Political Economy of Soil Erosion in Developing Countries*. London: Longman.
Blaikie, P. and H.C. Brookfield (eds). 1987. *Land Degradation and Society*. London: Methuen.
Bryant, R. 2015. "Reflecting on Political Ecology." In Raymond Bryant (ed.), *The International Handbook of Political Ecology*. London: Edward Elgar Publishing Co.
Callon, M. 1998. "An Essay on Framing and Overflowing: Economic Externalities Revisited by Sociology." In Michel Callon (ed.), *The Laws of the Markets*. Oxford: Blackwell.

Cheon, Chihyung, Seongeun Kim, Heewon Kim, and Miryang Kang. 2021. *Hoheup Kongdongche-Misemeonji, Corona 19, Pokyeome Eungdaphaneun Kwahakgwa Jeongchi(Breathing Community – Science and Politics responding PMs, Corona 19 and Heat wave)*. Seoul: Changbi Press.

Choi, Jinkyul, Rokjin J. Park, Hyung-Min Lee, Seungun Lee, Duseong S. Jo, Jaein I. Jeong, Daven K. Henze, Jung-Hun Woo, Soo-Jin Ban, Min-Do Lee, Cheol-Soo Lim, Mi-Kyung Park, Hye J. Shin, Seongju Cho, David Peterson, and Chang-Keun Song. 2019. "Impacts of Local vs. Transborder Emissions from Different Sectors on PM 2.5 Exposure in South Korea During the KORUS-AQ Campaign." *Atmospheric Environment* 203: 196–205.

Choi, Yeonjin. 2017. "The First Trial of PM against the Government of China." *Chosun Ilbo* (2017.4.6) (http://srchdb1.chosun.com/pdf/i_service/pdf_ReadBody.jsp?Y=2017&M=04&D=06&ID=2017040600121 http://cafe.naver.com/dustout) (in Korean).

Hwang, Inchan and Jongrak Paik. 2020. "Plan of Seoul Metropolitan City for Strengthening International Cooperation on PMs in Case of China." Policy Report of Seoul Research Institute (2020.12.14).

Hwang, Kihyun. 2018. "'Blood Weekend' . . . a Moving Image of the Largest-scale PM that Will Hit the Korean Peninsula this Weekend." *Life* (2018.11.9) (www.insight.co.kr/news/189930).

Jang, Jaeyeon. 2019. *Kongi paneun sahoie bahndaehanda (Against a Society that Sells Air—Air that Became a Product, Air that Became Fear, and a Korean Society that Reads through the Frame of Fine Dust)*. Seoul: Dongasia Publication.

Jung, Il-jun. 2012. "Jigujeok Wiheom Sahoiwa Segyesiminjui-Je 2cha Keundaesungeu Dayangseonggwa Sahoi Ironeui Segyesiminjeok Chunhwan" (Global Risk Society and Cosmopolitanism: The Variety of Second Modernization and the Cosmopolitan Turn of Social Theory). *Korean Journal of Sociology* 46(1): 200–205.

Kim, Yungwook, Hyunseung Lee, Youjin Jang, and Hyejin Lee. 2015. "How Does Media Construct Particulate Matter Risks?: A News Frame and Source Analysis on Particulate Matter Risks. *Korean Journal of Journalism & Communication Studies* 59(2): 121–154.

Korea Institute of Environment and Science. 2020. *Annals of Atmospheric Environment 2019*. Incheon: Korea Institute of Environment and Science.

Latour, B. 1997. "On Actor-Network Theory: A Few Clarifications Plus More Than a Few Complications" (http://cours.fse.ulaval.ca/edc65804/latour-clarifications.pdf).

Ministry of Environment. 2019a. *Whitebook of Environment, 2018*. Sejong: Ministry of Environment.

Ministry of Environment. 2019b. *Publication of the Report on the International Joint Research Project for Long-range Transborder Air in Northeast Asia*. Sejong: Ministry of Environment Press Release.

Park, Minhee. 2021. *Joonggook Dilemma (China Dilemma)*. Seoul: Hankyore Press.

Seoul Metropolitan Government. 2021. *Comparative Analysis of Air Quality Management in Beijing and Seoul*. Seoul Metropolitan Government.

Song-chae, Kyunghwa. 2019. "The Reason for the Worst PM. China or Korea. . . . We Can Check it Immediately." *Hankyoreh* (2019.3.4). (www.hani.co.kr/arti/society/environment/884503.html#csidxee5403364451ca89cca33caae2819f1).

Swyngedouw, E. 2015. "Depoliticized Environments and the Promises of the Anthropocene." In Raymond Bryant (ed.), *The International Handbook of Political Ecology*. London: Edward Elgar Publishing Co.

Whitehead, M., R. Jones, and M. Jones. 2007. *The Nature of the State: Excavating the Political Ecologies of the Modern State.* Oxford and New York: Oxford University Press.

Yeo, Minjoo and Yongpyo Kim. 2019. "Trends of the PM10 Concentrations and High PM10 Concentration Cases in Korea." *Journal of Korean Society for Atmospheric Environment* 35(2): 249–264.

Part II

Regional and transboundary air politics

4 China's transboundary pollutants and Taiwan's air politics

Paul Jobin, Shih-Hao Jheng, and Chee Wei Ying

4.1 Introduction

Echoing Ulrich Beck and Natan Sznaider (2006)'s critique of methodological nationalism, the editors of the present volume claim the need for a methodological cosmopolitanism (see also Chou and Liou 2012; Chou 2018). In this chapter, we embrace this approach as a requirement for tackling the issue of transboundary air pollutants. We believe, however, that national boundaries remain important constraints, and we need to understand to what extent air pollutants—either domestic or foreign—influence national politics, as well as how science and politics are entangled with one another.

Drawing on Actor Network Theory (Callon 1986; Latour 1987; Law and Hassard 1999), we look at air pollutants as perfect "actants" with an agency of their own, obviously disregarding national boundaries and potentially disrupting domestic politics. Although this approach is compatible with methodological cosmopolitanism, rather than proposing governance policy, the task we assign ourselves in this chapter consists more modestly in following air pollutants in all aspects, from their "long-range transport" from China to their intervention in Taiwan's domestic politics.

As Asuka highlights in his contribution for this book, in East Asia, transnational initiatives to monitor air pollutants are hampered by serious tensions over other diplomatic issues between Japan, South Korea, and China. Lee and Paik (2020, and their chapter for this volume) further show that asymmetric barriers seriously hamper negotiations between China and South Korea. In the case of China and Taiwan, the problem is even worse since the two countries have no diplomatic relations (this remark also applies to the case of Hong Kong; see Wong's chapter in this book). China does not even acknowledge the sovereignty of Taiwan, claiming that the island is a Chinese territory. Consequently, Taiwan is excluded from many international conferences and organizations, starting with the United Nations and its agencies, as well as from regional initiatives, such as the East Asian Network for the monitoring of acid rain (EANET) and the Long-range Transboundary Air Pollutants (LTP) launched between South Korea, China, and Japan (see chapters by Asuka, Lee and Paik). It is unlikely that Taiwan will have a chance to participate in these conferences in the near future.

In this context, despite many resources, it is extremely difficult for Taiwan to join a transnational monitoring of air pollution (or other transboundary

DOI: 10.4324/9781003211747-6

hazards, such as the COVID-19 pandemic). However, despite this radical asymmetry between Taiwan and China, since 2011, Taiwanese and Chinese specialists of air pollution have met twice each year. These meetings can be seen as an alternative diplomatic channel or an ersatz diplomacy.

Inspired by the notion of hydropolitics, Lee and Paik (2020, and their chapter in this volume) introduce the notion of "atmospheric politics" to focus on the efforts of governments, experts, and policymakers who deal with transboundary air pollution through transnational and multilateral frameworks of cooperation or conflict resolution. To borrow the title of Lidskog and Sundqvist (2011), those involved in atmospheric politics set their sights on "governing the air" across the borders of nations. Our study also includes these actors, but air pollutants that cross the borders constitute only one aspect of air pollution politics at large, and not necessarily the most important. Hence, we need to also conduct ethnographic observations of social movements, low-profile meetings, and the daily politics of air pollution. For that purpose, we find the notion of "air politics"—a neologism inspired as well by water politics (or hydropolitics)—more appropriate. We thus define air politics as a flexible concept that includes both the geopolitics and domestic politics of air pollution. In other words, air politics is similar to atmospheric politics but more "down-to-earth" (Latour 2017a), literally, at the level where ordinary people actually breathe.

In the first part of this chapter, we thus consider if the rounds of discussion on air pollution between Taiwan and China could present an alternative to state-to-state negotiations on transboundary hazards. We draw on Latour's (2017b) call for a renewed conception of geopolitics and diplomacy in our time of climate emergency, and on Middleton and Allouche's (2016) sub-notion of "hydro-diplomacy" to introduce that of "air-diplomacy." Moore (2018) considers the extent to which contentious hydropolitics inside China influences its relations with its transboundary riparian neighbors. In the second part of this chapter, we reverse the equation to examine just how much transboundary air pollutants coming from China influence domestic air politics in Taiwan. Given the fundamental importance of China in Taiwan's society and political economy (Wu JM and Liao 2015; Fell 2016; Wu NT 2016; Schubert et al. 2016; Wu JM 2020), as well as the concern over air pollution in Taiwan, we assume that the issue of transboundary air pollutants from China plays a role in citizen mobilizations and electoral campaigns. However, we find that although air pollution is an important element in local elections, instead of transboundary pollutants from China, anti-pollution activists emphasize the burden caused by heavy industries in south and central Taiwan, a view that has been confirmed by surveys.

4.2 Data and method

This chapter is based on a multi-site observation over a period of four years, starting in March 2016 and ending in November 2019. For the first part of this chapter addressing the issue of transboundary pollutants, our research includes interviews with ten air pollution experts (many of whom have contributed to writing reports for Taiwan EPA); attendance at three scientific conferences gathering Taiwanese and Chinese experts of air pollution; and the analysis of about thirty scientific articles and fifteen reports on air pollution from Taiwan EPA.

The second part, centered on domestic air politics in Taiwan, starts with an analysis of data from the Taiwan Social Change Survey (TSCS) and the Taiwan Social Image Survey, with an emphasis on a recent survey for which we contributed to preparing the questionnaire (Fu 2020).[1] This survey offers a precious basis for identifying how the population in Taiwan generally perceived transboundary air pollutants from China, whether they are an important matter of concern compared to domestic sources of air pollution, and how geographic and political preferences might influence the perception of these different sources, domestic and transboundary.

The following section further looks at these questions from qualitative observations we conducted at different sites and times, including street protests against air pollution in central Taiwan (Taichung and Yunlin), occasional public meetings with the EPA, and regular group meetings in Taipei with environmental activists from across the country and other scholars. Finally, we carried out a study of the 2018 local elections, which includes a big data analysis of online articles and comments over a one-year period.

The combination of these different kinds of data and methods allows for the issue to be grasped from complementary perspectives. The diversity of data forms a bundle of evidence, which allows us to confirm intuitions obtained from one approach, or conversely, helps us to temper judgments that otherwise would have been exaggerated. The result is both more nuanced and solid.

4.3 "Overseas pollutants" or the China factor of air pollution in Taiwan

According to an independent report based on data from NGOs and from governmental sources (IQAir 2020), out of a total of 98 countries, Taiwan ranks 58th for its average concentration of $PM_{2.5}$ (fine particulate matters up to 2.5 microns in size) as weighted by population, far behind China, which ranks 11th. The ranking is similar for their capital cities: despite considerable amelioration over the last five years, Beijing remains one of the ten most polluted (9th), while Taipei is 56th. Like Beijing, Shanghai and Chengdu have seen a sharp decrease in their concentrations of $PM_{2.5}$, but 48 Chinese cities still feature among the top 100 of the world's most polluted cities.

Because a good deal of China's air pollution finds its way to Taiwan, the quantity of air pollution emitted in China that travels over Taiwan has long been a growing matter of concern on the island. Reports from Taiwan EPA reflect this concern about transboundary pollutants. As we show in Table 4.1, in a total of fourteen reports on the concentration of $PM_{2.5}$, from 2011 to early 2020, the average share of foreign sources was about one-third (35.3 percent). There was no obvious evolution over the years; in the first report (based on data from 2007), the ratio of foreign sources was 37 percent, while in the last report (based on data from 2016), it was 38 percent; the ratio has thus remained stable, oscillating between approximately 20 and 40 percent.

Apart from one report clearly identifying "Mainland China" as the main source of transboundary pollutants, and another report referring to "East Asia," all of these reports used the expression "overseas" or, literally, "over the border" (*jingwai*).

As one of these authors explained:

> Our goal has been to clarify the impact of transboundary pollution on Taiwan. Using air quality modeling and excluding pollution sources from Taiwan, we look at long-range transports, not only from China but also from Korea, Japan, etc. . . . all over Asia. At the beginning, we were not so sure how important China's proportion was, then it became clear that it was the main foreign source.[2]

The research on "long-range transport" (LRT) of air pollutants dates back to the 1980s, when acid rain fell on Germany and the USA (on this early stage of research, see Asuka's chapter, and Lidskog and Sundqvist 2011). The "China factor" of this air pollution—as Interviewee A spontaneously called it—gradually became a topic in Taiwan in the 1990s, with young PhDs coming back from the USA. The first publications appeared in the early 2000s (e.g., Chang et al. 2000; Chen et al. 2002). They highlight that the dust carried from desert regions in northern China hits northern Taiwan during the fall and winter with various consequences for air pollution. For instance, Chou et al. (2004) show that long-range transported aerosols can sometimes contribute as much as 87% to the total quantity of PM_{10} in Taipei. During the winter and spring, there are also frequent episodes of elevated PM_{10} and $PM_{2.5}$ levels (Chou et al. 2007).

In addition to seasons and weather patterns, the share from China depends on the kind of pollutants observed—that is, $PM_{2.5}$, PM_{10}, sulfur oxide (SO_x), nitrogen oxide (NO_x), carbon monoxide (CO), dioxide (CO_2), and volatile organic compounds (VOCs). The location in Taiwan is also a crucial factor, the most affected area being the west coast from north to south. Lin et al. (2005) suggest, for instance, that 60 percent of CO in central Taiwan can be anthropogenic emissions from China.

More recent publications have confirmed these tendencies. For example, analyzing data from 1994 to 2010, Chen JP et al. (2015) found that foreign sources of PM_{10} carried by dust from northern China have generally been much higher than domestic emissions, with a seventeen-year mean of 57 percent. Although $PM_{2.5}$ behave differently, Chen TF et al. (2014) further show that on

Table 4.1 Transboundary share of $PM_{2.5}$ in Taiwan EPA reports (2011–2020)

Year of publication	Report title (all reports are in Chinese but also include an English title)	Main Author	Share and Appellation of Foreign Air Pollutants	Period of data collection
2011	A support to air quality modeling and control strategy (II)	Chang, Ken-Hui	Overseas (*jingwai*): 37%	2007
2013	Technical support for air quality model and assessment of air pollution control strategies (II)	Chang, Ken-Hui	Overseas: 37%	2006–2011
2014	Study on the spatial and temporal distribution characteristics of fine particulate matter ($PM_{2.5}$)	Yeh, Hui-Chung	Overseas: 37%	Not mentioned, but probably referring to the 2013 report.
	Technical support for air quality model and assessment of air pollution control strategies (III)	Chang, Ken-Hui	East Asia (*dongya*): 35%	2007
2015	Analysis of $PM_{2.5}$ reduction efficiency and development of its control program	Wu, Yee-Lin	Overseas: 30–50% (Average: 40%)	2007
	Integration of various departments to accelerate air quality improvement	Chen, Hsien-Heng (EPA)	Overseas: 27%	Not mentioned
	Clean Air Act (2015–2020)	Tsai, Hung-Te (EPA)	Overseas: 43.3%	Not mentioned
2016	Development of fine particle ($PM_{2.5}$) control strategy and assessment of their effects	Wu, Yee-Lin	Overseas: 30–50% (Average: 40%)	2007
	Project for consolidating the system of air quality model (I)	Chang, Ken-Hui	Overseas: 10–42% (Average: 26%)	2010
2017	Project of consolidating the system of air quality model (II)	Chang, Ken-Hui	China Mainland (*zhongguo dalu*): 5–39% (Average: 22%)	2013
	Current air quality and control strategies in Taiwan	Tsai, Hung-Te (EPA)	Overseas: 34–40% (Average: 37%)	2007–2010
2018	Project of consolidating the system of air quality model (III)	Chang, Ken-Hui	Overseas: 31%	2013
2020	Control measure cost analysis and health benefit of air quality control strategy	Lai, Hsin-Chih	Overseas: 21–67% (Average: 44%)	2013
	Establish domestic AERMOD model and air quality model validation system	Chang, Ken-Hui	Overseas: 38%	2016

an annual average, 36 percent of $PM_{2.5}$ levels are long-range transports from East Asia, of which China's "contribution"—as experts call it—ranges between 60 and 90 percent depending on the type of pollutants. In South Taiwan, East Asian long-range transports account for 32.9 percent on a yearly average (Lu et al. 2016).

Referring to one of the EPA reports (Table 4.1: Chang 2016 II), the Taiwan EPA Director for the Department of Air Quality Protection and Noise Control explained during an academic meeting:

> If we look at overseas emissions, a typical factor is the northeast monsoon. When the northeast monsoon goes southward, it slowly brings pollutants from Beijing to the Yangtze River Basin, then from the Yangtze River to Taiwan. The proportion that reaches Taiwan is about 1/6 of Beijing's. But this 1/6 affects many people in Taiwan, as it means the AQI turns to orange (over 100 or more). This is harmful to sensitive groups in the population (compared to red, which means harmful to all groups). During the winter, it can account for up to 40 per cent of total air pollution.[3]

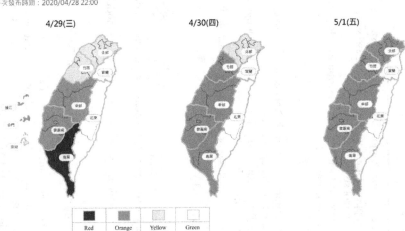

Figure 4.1 An example of a three-day forecast mapping of air quality, published by Taiwan EPA on April 28th, showing different alert levels: on April 29th, the southwest is red on the original (here in black), the central west orange (medium gray), and the north yellow (pale gray). On May 1st, the whole west coast is orange. The east coast remains at the safe green level (white) for all three days

Source: airtw.epa.gov.tw/CHT/Forecast/Forecast_3days.aspx

Figures 4.2 Seasonal contrasts of PM2.5: at left, during the winter (January 19, 2019, at 9 pm), the dominant winds (in arrows) blow from north to south.

Source: LASS, purbao.lass-net.org

Figures 4.3 Seasonal contrasts of PM2.5: in summertime (July 21, 2019, at noon), the dominant winds do the reverse, from south to north

Source: LASS, purbao.lass-net.org

But some researchers are dubious about China's impact on Taiwan, as one of them explains:

> I don't think China contributes that much to Taiwan's pollution. (. . .) EPA's assumption of a contribution up to one third or one quarter comes from modeling simulations. (. . .) Based on our observations, those long-range transport pollutants sometimes pass through at 2 or 3 km above the ground during some events. In other words, those pollutants did not reach the surface of Taiwan; they didn't affect air quality in Taiwan (. . .). Modelers think that what happens at 2–3 km above the ground is the same as on the surface. But they are different.[4]

Yet with the use of a laser-radar called Ladar, it is possible to conduct a "vertical profile" of air pollution, that is, measuring atmospheric aerosols at different altitudes.[5] In addition, as Edwards (2010) has shown in the case of scientific knowledge on global warming, data sampling could not be interpreted without models. Instead of blaming the use of model-based methods, the other researchers we met, therefore, emphasized the limitations of the data that are put into the model.

After the 2008 Beijing Olympics, the American embassy released its own Air Quality Index (from their own rooftop measurements), from lack of trust in the data published by the Chinese authorities. This episode provoked diplomatic tensions between the two countries. According to one Taiwanese researcher, official Chinese data became more reliable afterward.[6] However, as he further explains, the current monitoring grid in China can be as large as 625 square kilometers, whereas, in Taiwan, the grid is one square kilometer, which allows for greater precision.[7] This difference means that compared to Taiwan, in China it is more difficult to identify emission sources such as industrial zones or highways.

4.4   Air-diplomacy and cross-strait politics

In their study on the long-range transport of $PM_{2.5}$ from Asia and China to Taiwan, Chen TF et al. (2014) concluded: "The only logical alternative is to negotiate with China in reducing its SO_x content, which would reduce the quantity of direct SO_4^{2-} transport." We thus now consider an example of such negotiations. The reader should keep in mind the very particular context that characterizes relations between Taiwan and China, usually called "cross-strait relations" (*liang'an guanxi*), a euphemism to avoid the vocabulary that goes with normal state-to-state diplomacy (i.e., expressions like "Taiwan and China," which could imply equal status).

Since 2011, through a professional association (Taiwan Environmental Management Association), Taiwanese air pollution scholars have launched meetings with Chinese counterparts from universities such as Tsinghua and Beida as well as regional universities. The meetings have been held every year, alternating between China and Taiwan, with two meetings per year, one devoted to academic exchange, the other to policy and administrative talks. The latter has had the participation of Taiwan EPA executives and representatives from China's Ministry of Ecology and Environment.[8] In 2020 and 2021, these meetings were suspended due to the COVID-19 pandemic.[9]

In their study of hydropolitics in the Lancang-Mekong basin, Middleton and Allouche (2016) introduce the sub-notion of "hydro-diplomacy" to describe the diplomatic competition between China, Thailand, and Vietnam for their numerous programs of dam construction. As they show, such hydro-diplomacy entails a subtle mix of conflict and cooperation. In the case of China, the goal is not only to control the river's headwaters but also to expel the USA and Japan from mainland Southeast Asia, which Chinese authorities consider their backyard. In comparison, the cross-strait meetings between Chinese and Taiwanese specialists in air pollution are not contentious. Indeed, they carefully avoid any potentially antagonistic discussion.

These cross-strait talks went smoothly during the presidency of the Kuomintang (KMT)'s Ma Ying-jeou in Taiwan, who implemented several pro-China policies. However, as two scientists and a top EPA executive have confessed, the roundtable discussions between state officers were suspended after the election of Democratic Progressive Party Tsai Ying-wen in 2016, who refused to endorse the so-called "1992 consensus."[10] Beijing was not ready to give Tsai's government any opportunity to meet with Chinese officials, even in such usually non-confrontational meetings. As a result, the only channel left for discussion was academic exchange.[11] Moreover, in contrast with most academic conferences, these meetings must keep a low profile; they are not announced publicly and participation is closely monitored.

In general, our interviewees emphasized the positive aspects of these academic exchanges. Even under repeated Chinese military threats against Taiwan, scientists can keep on doing science as usual. For a Taiwanese researcher who has co-published with Chinese colleagues from top-rank universities:

> It would be going too far to say that academic exchange has replaced state diplomacy, but the Chinese stress personal relations, so it is good that we have developed those links with Chinese colleagues. Over the years, we have become friends and we stick to scientific research. We avoid political topics.[12]

Though admitting that it is sometimes difficult to make a clear-cut break between science and politics, this researcher emphasizes that academic exchanges still provide precious informal occasions for scientific discussion and cooperation. But the issue remains sensitive, especially vis-à-vis the accuracy and precision of data, a problem at the heart of the tensions with the USA during the 2008 Beijing Olympics and, more recently, the polemics over the origin of COVID-19.

Still, it is not clear when the issue of transboundary pollutants from China can be concretely addressed. For example, toward the end of one of these meetings, the following exchange took place:

> A Taiwanese participant: "The proportion of air pollutants from China has received a lot of attention in Taiwan. Recently, the news said that most of

Taiwan's air pollution was caused by the northeast monsoon from mainland China. [. . .] If the media reports erroneous statements about the issue, it is better that you clarify it, otherwise it will deepen misunderstanding on the two sides of the strait."

A Chinese participant: "The relative impact of this pollution is an objective fact. We are all in the same atmosphere, but the impact of air pollution is measured by scientific research. It is specific to each pollution process and the situation of different cities varies a lot, so I think cross-strait cooperation on the issue is about studying the atmosphere. Regarding the state of the air, either on a large or small scale, we are very concerned about the interaction of pollution on various scales, like for example in the case of our efforts to reduce emissions in our region. But all this can be discussed."[13]

Indeed, previous to that conference, a veteran Taiwanese researcher had warned us that it was difficult to address the issue of transboundary pollutants during these meetings:

"This is a sensitive issue. If we say that one third of air pollution in Taiwan comes from China, our Chinese colleagues might become nervous. They might say that Taiwan also has an impact on the Mainland. If they admitted that China's impact on Taiwan is about one third, it would mean that China should take responsibility, right? So we'd better not talk about this."[14]

The executive in charge of air protection at Taiwan EPA concluded philosophically:

"With such a high share of overseas emissions, between 30 and 40 per cent, what can Taiwan do? Not much. But we can manage the remaining 60 to 70 per cent. This is our main duty."[15]

Echoing these data, from time to time, Taiwanese media has made a sensational topic of air pollution from China. However, the population seems to react with only moderate interest, and variables such as political orientation may create perception gaps. In the remainder of this chapter, we will thus examine the impact of air pollutants from China on domestic politics in Taiwan. For this, we will mobilize different methodologies and sets of data, starting with surveys.

4.5   Public perception of air pollution

In this section, we borrow data from surveys conducted by the Institute of Sociology at Academia Sinica for the Taiwan Social Change Survey (TSCS) and the Taiwan Social Image Survey (TICS), which offer a rich set of data.

According to a Social Change Survey conducted in 2010 devoted to environmental issues (Chang 2016), air pollution (31.9 percent) was considered the most important problem, ahead of climate change (19.1 percent), depletion of natural resources (12.6 percent), water quality (11.6 percent), pesticides and chemical fertilizers (11 percent), drought (5.3 percent), nuclear waste (2.3 percent), and genetically modified foods (1.3 percent).[16] Air pollution from industries was seen as dangerous or very dangerous for the environment (85.7 percent) against 55.6 percent in the case of car emissions.[17]

This high level of concern over air pollution was confirmed by a Social Image Survey conducted at the end of 2017 (Yang 2021), with 82.2 percent of respondents feeling worried or very worried about it;[18] and 77.9 percent estimating that air pollution has a direct impact on their daily life.[19] For the first time, the topic of air pollutants from China was introduced with this question: "Who should be held accountable for air pollution in Taiwan?"; 39.3 percent of respondents put "companies" first, followed by the "central government" (27%), while "Mainland China" came in third with a relatively low 14.7 percent, slightly above the responsibility attributed to "local governments" (14.1 percent). Answers to this question suggested that Taiwanese do not seem to feel overwhelmed by a "Chinese invasion" of toxic air. But this issue is worth a closer examination.

Following on a seminal edition in 2013, a Social Change Survey conducted in 2019 was devoted to risk perception (Fu 2020). As two authors of the present chapter (Jobin and Jheng) participated in the preparation of the questionnaire, we proposed including questions related to air pollution and the problem of foreign sources. The remainder of this section presents an analysis of these results.

A series of nine questions dealt with various matters of concern, from natural to technological hazards, such as climate change, nuclear accidents, GMOs or food contamination from pesticide residue.[20] Air pollution ranged among the main concerns, with 46.4 percent of people worried and 29.1 percent very worried, for a total of 75.5 percent, compared to 64.7 percent for the consequences of global warming and 50.3 percent concerning the risk of a nuclear disaster.[21]

When asked about the main sources of air pollution, the respondents had the choice between five options in the following order: "overseas pollutants (such as sandstorms from China)," "thermal power plants," "big factories," "diesel transportation (e.g., trucks, airplanes, ships)," and "cars and scooters." Only 16 percent of interviewees estimated that overseas pollutants were the main cause of air pollution, a result slightly above the 14 percent found in the 2017 survey (*supra*). Although overseas pollutants were listed first in the questionnaire, perception of this risk fell behind that of cars and scooters, power plants, and big factories (respectively 19.1, 23, and 26.3 percent).[22]

As we saw in the previous section, measurements of air pollution in Taiwan show important seasonal and geographic variations (see Figures 4.1–3 *supra* and 4.4 *infra*). So as to check how this uneven distribution might influence risk perception, we divided survey responses into four regions according to the interviewees' place of residency (and site of interview).

As summarized in Table 4.2, from north to south on the western side of Taiwan, approximately the same proportion of respondents (between 16 and 18 percent) identify overseas pollutants as the main source of air pollution, in sharp contrast with residents from the east side (less than 7 percent). Residents from the central west put thermal plants in first position (near 30 percent), while 35 percent of southwestern residents point at big factories, reflecting the geographic concentrations of these industries (see Figure 4.4 *infra*).

In addition to place of residency, political tendencies might constitute another variable that significantly influences the perception of this issue. We thus distribute the respondents according to their vote during the local elections of November 2018—for mayors and delegates of cities and counties—into three groups: the pan-blue voters (Kuomintang, New Party, and People First Party) who tend to look on China favorably, the pan-green voters (the Democratic Progressive Party/DPP, and the New Power Party) who tend be wary of China, and a third group comprising those whose party affiliation is unknown as well as those who did not vote (Table 4.3).

Table 4.2 Perception of air pollution sources among voters in the local elections of November 2018

Air pollution sources Region	Overseas	Thermal plants	Factories	Diesel	Cars and scooters	Total
North West	157	213	206	70	210	856
	18.34	24.88	24.07	8.18	24.53	100.00
Central West	58	106	100	36	57	357
	16.25	29.69	28.01	10.08	15.97	100.00
South West	83	83	156	51	70	443
	18.74	18.74	35.21	11.51	15.80	100.00
East	7	36	31	8	24	106
	6.60	33.96	29.25	7.55	22.61	100.00
Total	305	438	493	165	361	1,762
	17.31	24.86	27.98	9.36	20.49	100.00

Pearson chi2(12) = 54.3323 Pr = 0.000 (below 0.05 is significant)

Source: Fu 2020 (current residency and question G2)

Table 4.3 Respondents' votes in the local elections of November 2018

Vote	Freq.	Percent
Pan-blue (KMT, NP, PFP)	581	38.84
Pan-green (DPP, NPP)	453	30.28
Others	462	30.88
Total	1,496	100.00

Source: Fu 2020 (question K11)

We then cross these voting behaviors with the respondents' levels of concern for air pollution (Table 4.4). There is no major discrepancy between the three groups, with plan-blue and pan-green voters very close to one another (respectively 77.3 and 79.2 percent of worried and very worried), and the third group of non-voters and voters of unknown affiliation following not far behind (with a total of 69.7 percent of worried and very worried). These results suggest that political preference has little or no correlation to concern regarding air pollution.

However, when we combine the respondents' political votes with their perception of air pollution sources (Table 4.5), the results show a significant difference between pan-blue and pan-green voters. While the first group underlines the responsibility of thermal plants (35.1 percent of respondents) and downplays overseas pollutants (12.8 percent), responses from the latter tend to the reverse, with 27.6 percent of pan-green voters pointing more to overseas pollutants, rather than to thermal plants (13.7 percent).

Table 4.4 Concern about air pollution among voters in the local elections of November 2018

Air pollution risk Vote	Very worried	Worried	Don't care	Not worried	Absolutely not	Don't know	Total
Pan blue	178	271	44	73	15	0	581
(KMT, NP, PFP)	30.64	46.64	7.57	12.56	2.58	0	100
Pan green	147	212	40	47	6	1	453
(DPP, NPP)	32.45	46.8	8.83	10.38	1.32	0.22	100
Else	120	202	44	77	19	0	462
	25.97	43.72	9.52	16.67	4.11	0	100
Total	445	685	128	197	40	1	1,496
	29.75	45.79	8.56	13.17	2.67	0.07	100

Source: Fu 2020 (questions B2 and K11)

Table 4.5 Perception of air pollution sources among voters in the local elections of November 2018

Air pollution sources Vote	Overseas	Thermal plants	Factories	Diesel	Cars and scooters	Total
Pan-blue (KMT, NP, PFP)	72	197	126	66	100	561
	12.83	35.12	22.46	11.76	17.83	100.00
Pan-green (DPP, NPP)	118	59	117	43	91	428
	27.57	13.79	27.34	10.05	21.26	100.00
Others	65	110	132	27	99	433
	15.01	25.40	30.48	6.24	22.86	100.00
Total	255	366	375	136	290	1,422
	17.93	25.74	26.37	9.56	20.39	100.00

Source: Fu 2020 (questions G2 and K11)

Pearson chi2(8) = 92.7190 Pr = 0.000

A quick interpretation of these results would be that pan-green voters tend to harbor apprehensions toward China, and hence are likely to blame China for air pollution, whereas pan-blue voters are more likely to show the opposite tendency. Although this view might be more or less correct, we nevertheless need to consider more carefully how domestic politics at the national and local levels entangle with geographic discrepancies, particularly given the uneven distribution of thermal plants and heavy industries between the north and the south of the island. The next section attempts to further analyze this problem from a different angle, based on qualitative observations of the protest movement against air pollution in Taiwan.

4.6 Popular protest: "one sky, but two Taiwans"

In Taiwan, the movement against air pollution has carried on a legacy of early battles against the hazards of heavy industry, in places like the Houchin and Linyuan districts in Kaohsiung (Ho 2005; Lu 2016). Another hotspot has been the fence-line communities around the petrochemical zone of the Sixth Naphtha Cracker (controlled by Formosa Plastics) in Yunlin and Changhua counties. In 2010–2011, a project to extend the zone by Kuokuang Petrochemical triggered a nationwide opposition movement (Ho 2014b; Grano 2015). Furthermore, the dramatic increase in cancers, less than ten years after the cracker's start of operations (in 1998), has attracted the attention of public health researchers, lawyers, and activists (Jobin 2021).

Around 2013, the mobilization against air pollution benefitted from a new impulse from several initiatives of citizen science in central and south Taiwan (Tu 2019). In 2015, the Chinese documentary "Under the Dome" (*qiongding zhi xia*) provided additional momentum to the movement in Taiwan, with a new wave of actors: urban middle-class people worried about the frequent smog and the consequences for their children's health from regular exposure to carcinogenic pollutants like $PM_{2.5}$ (Chen and Ho 2017).

Compared to south and central Taiwan, the population of north Taiwan's urban areas (Taipei, Taoyuan, Hsinchu) has been less exposed to air pollution from heavy industry. But they are exposed to intense pollution from cars, trucks, scooters, and motorcycles, as reflected in Table 4.2 (Taiwan has around 13 million motorbikes for a population of 23 million, one of the highest ratios in the world). In wintertime, the long-range transport of air pollutants from China only makes things worse.

Nonetheless, the problem of air pollution has been resented with more bitterness in the south (around Kaohsiung City), and the central west area (around Taichung City) and Yunlin County, where heavy industries like steel mills, petrochemical plants, oil refineries, and thermal plants are concentrated (see Figure 4.4).

These sites have more often been exposed to red alerts and "purple explosions" (*zibao*) of air pollutants than other areas of the island.[24] In Taiwan, a structural element of air politics is thus the division between the north and

Figure 4.4 The main sites of Taiwan's heavy industry: Kaohsiung, Yunlin, and Taichung.[23]

south of the island. This opposition, more generally speaking, has been marked by sociological differences, with the south experiencing higher rates of unemployment, lower levels of education, and steady migration to the north.

The concern over air pollution in south and central Taiwan has attracted the solidarity of Taipei-based scholars and environmental organizations. For instance, on February 19, 2017, Nobel Prize chemistry laureate Lee Yuan-Tseh and other celebrities joined thousands of people for a demonstration in Taichung City under the banner of "one sky, but two Taiwans" (*yige tiankong, liangge Taiwan*), a slogan that decries the geographical inequalities and implicitly longs for an idealized national unity (see Figures 4.5–4.7). The protesters stressed the responsibility of heavy industry and the central government's lack of commitment to solving the pollution problem in south and central Taiwan.

Figure 4.5 Demonstration against air pollution in the shape of Taiwan with the banner "One sky, but two Taiwans," in Taichung, February 19, 2017.

Source: (Photo: Apple Daily, Taiwan)

Figures 4.6–7 Demonstration against air pollution in Taichung, February 19, 2017. At the top, personalities on stage, including Nobel Prize winner Lee Yuan-Tseh and public health scientists from National Taiwan University (in Taipei). At the bottom, a schoolteacher with his students wearing paper hats in the shape of smokestacks.

Source: (Photos: P. Jobin)

The chronic episodes of red and purple alerts have been the topic of heated disputes among specialists. Researchers working with the EPA tend to emphasize the diversity of sources (i.e., not only industries, but also vehicles, merchant ships, etc.), as well as seasonal and geographic factors. These factors explain why even towns in non-industrialized areas in central Taiwan, like Chiayi and Nantou counties, have had bouts of record pollution, inspiring grassroots initiatives to reduce emissions from motorcycles or those resulting from burning plant waste and—in a common religious practice—incense and paper money (Liu 2019). However, for other experts and environmental organizations, the main culprits remain the heavy industries that frequently disregard emissions standards and use camouflage to present positive data. Firms like Formosa Plastics and Taipower have shown a special gift for this exercise. Another corporate tactic is to shift the blame onto vehicle emissions and long-range transports from China, but this argument does not really deter the populace's anger toward industrial polluters.

According to Taiwan EPA, the level of $PM_{2.5}$ has been gradually decreasing over the last ten years throughout the country, although the south (the region of Kaohsiung) and the central west (Taichung) continue to bear the brunt of it (see Figure 4.8). However, independent data-sharing platforms of citizen science have increased awareness of the risk.[25] From time to time, discrepancies between these data sets spark debate on social media, with concerned citizens or environmental groups accusing the EPA of falsifying its computations (Jheng 2019).

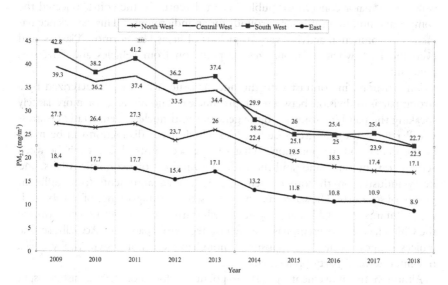

Figure 4.8 $PM_{2.5}$ in Taiwan (2009–2018)
Source: Data from Taiwan EPA

Scientists and grassroots mobilizations of citizen scientists have also played an important role in this movement (Tu 2019). Two researchers, in particular, have challenged the government and heavy industries, stressing that the EPA has been overestimating the proportion of air pollutants attributed to vehicles and foreign sources. Their voice has been influential within the citizen movements. One is Chang-chuan Chan, an epidemiologist specializing in air pollution, graduate of Harvard University, and dean of the School of Public Health at National Taiwan University (in Taipei); his research has highlighted the impact of industrial pollution in south and central Taiwan (e.g., Shie and Chan 2013; Yuan et al. 2018; see also Jobin 2021).

The other researcher is Tsuang Ben-Jei, professor of environmental science at National Chung Hsing University, in Taichung. His research has shown the dispersion trajectories of air pollutants (SO_2, NO_2, and $PM_{2.5}$) according to winds and other seasonal meteorological variations, citing in particular the emissions coming from the Sixth Naphtha Cracker petrochemical zone in Yunlin and other industrial complexes like the Taipower thermal plant in Taichung. During the mobilization against the Kuokuang project (*supra*), Tsuang estimated a possible excess of 1,356 deaths per year due to an increase in cancer and cardiovascular disease among residents, if the plant were to be built.[26] In 2012, Formosa Plastics sued him for defamation, both in criminal and civil court, on the pretext that Tsuang's public declarations (to the media, at academic conferences, in EPA meetings, etc.) were not based on published results. The company requested damages of NT$40 million (US$ 1.3 million). Tsuang received the support of lawyers, environmental NGOs, students and academics, all shocked by this threat to academic freedom and the right to debate a highly controversial issue and a major concern for public health. Eventually, the courts rejected the complaints, finding insufficient grounds for an indictment.[27] This experience has made environmental activists very suspicious of firms like Formosa Plastics, and even the EPA when it emphasizes air pollution from vehicles and "overseas pollutants."

Furthermore, in contrast with the nuclear issue, which is characterized by a strong partisan division between the pro-nuclear Kuomintang (or more largely speaking the pan-blue camp) and the generally anti-nuclear Democratic Progressive Party (or the pan-green camp), no such clear-cut distinction can be made in the debate on air pollution. What can be seen instead is a clash between activists on the one hand, and the EPA, the Ministry of Economic Affairs, and heavy industry on the other, each side backed by arguments from different groups of scientists. The scientists who underscore the importance of transboundary pollutants from China are not necessarily pan-green, and those who criticize the China factor as overemphasized are not necessarily pan-blue. Actually, several indices suggest rather the opposite, although we did not systematically check our interviewees' party preference.

Although the movement against air pollution does not have a distinct split between the pan-green and pan-blue camps, it overlaps with a constant factor in Taiwanese politics: south Taiwan—Kaohsiung and Tainan in particular—generally supports the DPP, while the KMT enjoys more support in north Taiwan, particularly in Taipei and Hsinchu. However, in the next sections, we will see how air politics facilitated the first success of the KMT in environmental politics in south and central Taiwan.

4.7 Political uses of air pollution

In December 2017, Tsai's government briefly attempted to spotlight air pollutants from China as the main source of air pollution, but this argument was immediately disproved (Chan 2017; Yin 2017; see also Tsuang 2018). Air politics, as it turned out, provided the KMT a much greater opportunity.

In the past, referenda on environmental matters were initiated by local communities opposed to incinerator or waste storage, and by the anti-nuclear movement with the support of the DPP (Ho 2006: 327–336). It was, however, difficult to stir up enough attention or sufficient voter support at the national level; and at the local level, powerful groups with financial interests could divert a referendum away from its initial goal of implementing grassroots democracy (Ho 2006: 340–343). A December 2017 reform of the Referendum Act rendered it far easier to introduce questions for public approval.

During the course of 2018, numerous referendum questions were proposed for inclusion on the ballot for the November municipal elections. The end result was a mishmash of ten questions about the best energy choices, food safety, national identity, and LGBT rights. The questions over energy choices were related to the issue of air pollution. One initiative drew particular attention. In March 2018, nuclear activist Huang Shih-hsiu, who had served as an adviser to former KMT chairwoman Hung Hsiu-chu, came up with a proposal to stop the DPP's plans to phase out nuclear energy. Under the slogan "green for nuclear" (*yihe yanglü*), Huang promoted nuclear energy as a means of reducing the air pollution from coal-burning thermal plants and spearheaded efforts to bring the matter to a public vote in a referendum. The KMT jumped at the chance, and Huang's bid to ensure the maintenance of nuclear energy eventually made the list, as did two KMT initiatives to cut coal-based energy production.

The very limited time between the approval of the referendum questions and election day made it very difficult for environmental groups to fight back with rational arguments on technical issues like energy and air pollution.[28] The presence of such hot-button topics as same-sex marriage further marginalized these matters. Moreover, the multi-question referendum ultimately served as a wholesale denunciation of the DPP and the Tsai government and contributed to the election of KMT candidates. Among them were Han Kuo-yu in Kaohsiung and Lu Shiow-yen in Taichung. Both of them made reducing air pollution a campaign pledge. In Taiwan, issues such as air pollution are usually neglected during presidential and legislative elections; instead, national identity, Taiwan's status, and cross-strait policy play the crucial role (Wu JM and Liao 2015; Fell 2016; Wu NT 2016; Schubert et al. 2016; Wu JM 2020). But when it comes to local elections, the China factor is not the main issue: other topics, such as economics or the environment, generally have a bigger influence, and the local elections of November 2018 were no exception (Batto 2019).

To get a proximate idea of the possible contribution of air pollution in the local elections of November 2018, we conducted a big data analysis of a large number of posts on social networks such as Facebook, as well as online articles and comments, using a selection of keywords such as the names of the main candidates and a list of topics that made the headlines during the campaign, including air pollution.[29]

The publication timeline of articles related to "air pollution" (see Figure 4.9) reveals three peaks. The first was at the end of June, around the vote for the reform of the Clean Air Act on June 27. The second came on September 20—this time with a larger number of comments (approximately 28,000 comments on 550 articles)—when environmental groups, as well as New Taipei City candidate Hou You-yi, opposed the replacement of coal-burning units at the Shen'ao thermal plant near Taipei. The last peak was on November 8, with discussions on air pollutants from China and the child victims of the Yunlin petrochemical zone (see Tu Wenling's chapter in this book).

Figure 4.10 further presents the results of the main keywords correlated with air pollution (with two different sinograms for the Chinese character *wu* "pollution"). The chart on the left—made with the software InfoMiner—can only take account of the online articles, not their related comments. The results included a significant number of articles for KMT (2,758) and DPP (2,281), followed by Taichung (2,272) and Kaohsiung (1,708), as well as the names of three candidates for the mayoral elections in Taichung (Lu Shiow-yen and Lin Chia-lung, respectively, 2,281 and 2,225) and Kaohsiung (Han Kuo-yu, 982), but no other cities or candidates. This tends to confirm that the problem of air pollution gathered a certain importance during the campaign for the elections in Taichung and Kaohsiung. Lu was running against Lin, the incumbent DPP mayor of Taichung; the name of Chen Chi-mai, the DPP incumbent mayor for Kaohsiung did not appear among the results, but it does appear when combining "air pollution" and "Kaohsiung" (see Figure 4.11), although with a limited occurrence of 489 articles.

Han Kuo-yu's sudden popularity just three months ahead of the elections stirred lots of comments and rumors about possible underground support from China to back his campaign. Han's race for control of Kaohsiung City was certainly the most striking phenomenon of the local elections of 2018. With his sweeping statements and provocative out-of-the-box ideas, Han became a political phenomenon all over Taiwan. Both the local pan-green media and foreign liberal media presented him as an archetypal populist candidate. Although Han claimed that he would clear the city of air pollution, he nevertheless promised to maintain the presence of heavy industry, including thermal plants. He not only won the mayoralty of Kaohsiung, a city many considered a DPP hotbed, but he was also later selected as the KMT candidate for the 2020 presidential election. However, not only did he fail to win the presidential

Figure 4.9 Time of publication of online articles and comments correlated with "air pollution" (articles in bullets, comments in bars).

Source: Data source: InfoMiner

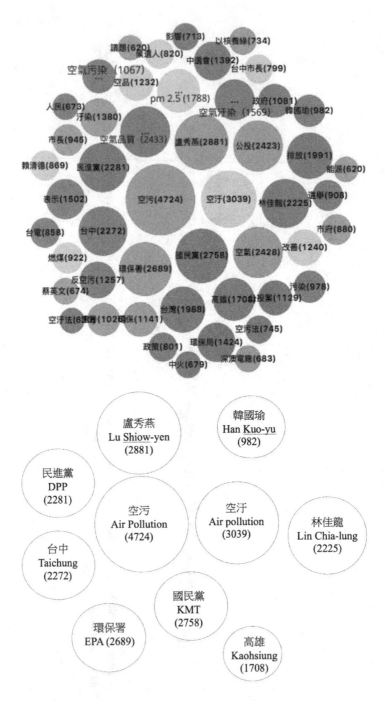

Figure 4.10 Online articles correlated with "air pollution" between June 1, 2018 and December 31, 2018. At the bottom is a selection of the keywords most relevant for the discussion here.

Source: Data: InfoMiner

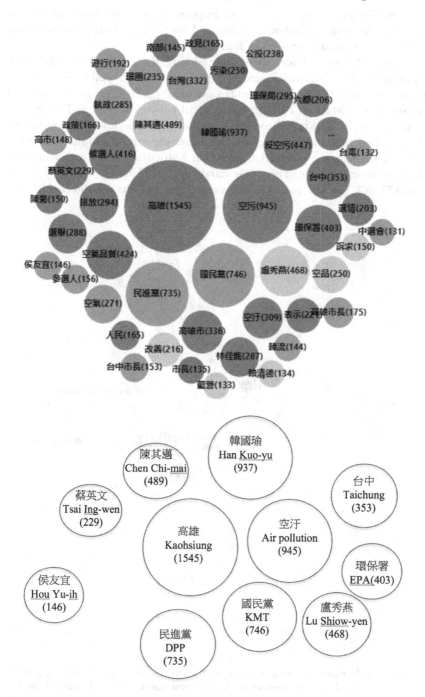

Figure 4.11 Online articles correlated with "air pollution" and "Kaohsiung"
Source: Data: InfoMiner

election, Kaohsiung residents were so disappointed with his breach of promise to stay in his post as Kaohsiung mayor and his frivolous style of management that he was eventually recalled in June 2020.

Figure 4.12 aims at identifying how, during the campaign for the 2018 elections, various hot topics—such as discussions on air pollution or LGBT rights—connected with opinion leaders on social media. This analysis is based on articles published on PTT, a popular social network limited to Taiwan. Our focus here was Han Kuo-yu and the locations of opinion leaders (identified by their PTT identification in English): the bigger their name and the thicker the row, the stronger the connection with Han Kuo-yu and other topics, and the stronger their influence. Three IDs stand out: "Weikitten," "Tiger Lily" and "mark2165." The first two (Weikitten and Tiger Lily) have a direct thick link

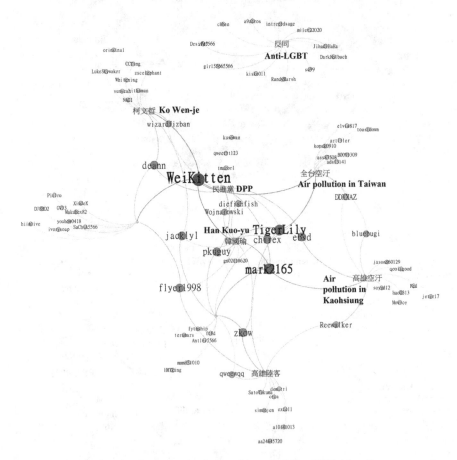

Figure 4.12 Network of opinion leaders and keywords on PTT (the chart was made on 2 modes analysis with the application Gephi, developed by the Medialab of SciencePo, Paris).

Source: Data: InfoMiner

with "air pollution in Taiwan," while mark2165 is strongly correlated with "air pollution in Kaohsiung."

These results are congruent with another big data analysis (Chiu 2019), which found that "mark2165" was a key ID for stirring up positive discussions on Han Kuo-yu; furthermore, 95 percent of the messages from that ID were posted in the PTT category of "Hate Politics" (*zheng heiban*, which we feel is better translated as "The Dark Side of Politics"), with nearly 70 percent of the content related to Han Kuo-yu, Kaohsiung, and Chen Chi-mai, with generally positive comments on Han. Our analysis further suggests that air pollution was among the topics which served as a pretext for this platform of "hate politics," thus contributing to the electoral success of Han Kuo-yu at the local elections of November 2018. With the exception of a few comments, air pollution from China was not a topic of these online discussions.

4.8 Conclusion

Unlike virus carriers, who can be checked and then tracked upon crossing borders—as Taiwan and South Korea have, to a large degree, successfully demonstrated in the case of COVID-19—air pollutants can spread their harmful effects freely with no respect for national boundaries. So while Taiwan has managed to overcome the difficulties posed by its unique geopolitical status—excluded from the WHO, the United Nations, and other international institutions at China's insistence—during the current pandemic, the problem of air pollution from China cannot be so easily handled. Still, despite the complex and adversarial relationship, Chinese and Taiwanese air pollution specialists have been able to conduct cross-strait meetings every year since 2011. Unfortunately, since the election of Tsai Ing-wen, China has increased its pressure on Taiwan, so that these meetings no longer openly address the issue of transboundary pollutants; their participants practice self-censorship, having deemed the subject too "sensitive" to be broached.

As discussed by Lee and Paik (2020, and in their chapter for this book), asymmetric power relations remain a serious problem for China's immediate neighbors. The authors emphasize in particular the asymmetry of domestic pressure. Whereas South Koreans urge their government to settle with China regarding the problem of air pollution (including transboundary PM), Chinese citizens may complain about domestic pollution in China but follow state media in arguing that South Korea generates its own pollution (Lee and Paik 2020: 133). The difference in political regimes—authoritarian in the case of China, democratic in the case of South Korea—no doubt contributes to nurturing different popular reactions and expressions of green nationalism.

The asymmetry of power with China is even worse in the case of Taiwan, and cross-strait issues exert a tremendous influence in the latter's domestic politics, in particular during presidential and legislative elections. Moreover, according to Taiwan EPA and multiple studies, transboundary pollutants from China account for around one-third of air pollution in Taiwan. Logically, the

China factor should therefore play an important role in domestic air politics. However, some researchers and anti-pollution activists challenge these results and stress instead the burden caused by heavy industries in south and central Taiwan. Finally, although air politics has an impact during local elections, as we have seen in the case of the 2018 elections, the role of transboundary pollutants from China has remained relatively marginal, almost a non-issue, which might explain local and central governments' lack of motivation to consider air diplomacy a priority issue.

Besides, if the confinement measures to prevent COVID-19 have had the beneficial effect of reducing air pollution in China, this lasts only as long as the confinement is maintained. Moreover, early results from research conducted in the USA (Friedman 2020) have found that air pollution increases mortality among people infected by COVID-19. These results were congruent with previous findings on the correlation between air pollution and SARS in China (Cui et al. 2003). Although the World Health Organization (WHO) remains cautious about the link between fine particles and the spread of the virus, these studies suggest that concern over COVID-19 (and in particular its variants) could put transboundary air pollutants from China back on the agenda of air politics in Taiwan. But as we show in this chapter, such "Chinese invaders" will not necessarily create a strong nationalistic response as was observed by Lee and Paik in the case of air politics in South Korea before COVID-19.

Acknowledgments to the editors and two anonymous reviewers for their stimulating remarks on the first manuscript, to Rebecca Fite for her fine-grained proofreading of the first and second manuscripts, and to Lin-Ru Hsu for her careful handling of the final manuscript.

Notes

1 The Taiwan Social Change Surveys (TSCS) and the Taiwan Social Image Surveys (TICS) have been conducted since the early 1980s by the Institute of Sociology of Academia Sinica on cohorts of approximately two thousand people through direct interviews (of around one hour each). In addition to around thirty questions on personal data (sex, age, education, profession, kinship, ethnic group, political tendency, etc.), each questionnaire (in Chinese) includes around one hundred questions on two specific topics which vary every year. While Social Change Surveys are conducted on specific topics once every five or ten years, Social Image Surveys are conducted every year and include a larger range of current issues.
2 Air pollution scientist C; interview in Taipei, 17 July 2017.
3 Oral presentation at Academia Sinica (AS), Taipei, 22 May 2017, with the participation of AS President.
4 Air pollution scientist D; mail communication, 8 July 2017, informal oral communication in Taipei, 25 July 2017, and in-depth interview, 17 July 2018.
5 Air pollution scientist E; interview in Taipei, 19 July 2017.
6 Air pollution scientist C; interview in Taipei, 17 July 2017.
7 Idem.
8 Air pollution scientist G; interview in Taipei, 28 September 2017.
9 Mail exchange with the organizers, July 2021.

10   The term was coined by the KMT to mean that China and Taiwan have agreed that there is only one "China," with each side having its own interpretation of what "China" means and who is its legitimate ruler (the Republic of China officially ruling the island, or the People's Republic of China that governs on the continent).

11   Idem; interview with scientist C (Taipei, August 2019); and discussion with the Taiwan EPA Minister, Taipei, 24 May 2017.

12   Air pollution scientist C; interview in Taipei, 17 July 2017.

13   14 December 2017, National Yunlin University of Science and Technology.

14   Air pollution scientist E; interview in Taipei, 19 July 2017.

15   Oral presentation at Academia Sinica (AS), Taipei, 22 May 2017, with the participation of AS President.

16   Chang 2016 (TSCS 2010), question 20a "What do you think is the most important environmental issue for Taiwan as a whole?" (Here *et infra* our translation from Chinese.)

17   Chang 2016, question 27a "Generally speaking, do you think air pollution caused by cars is dangerous to the environment?" and 27b "Generally speaking, do you think air pollution caused by industry is dangerous to the environment?

18   Yang 2021 (TSIS 2017), question 5, "Do you think Taiwan's air pollution is serious?"

19   Yang 2021, question 6, "In your daily life, are you often, sometimes, rarely or not affected by air pollution?"

20   Fu 2020 (TSCS 2019), question B2a-I based on the pattern "Do you worry the following problem may affect you or your family?"

21   The previous survey of 2013 did not include a question on air pollution, but global warming and nuclear disasters obtained higher scores, respectively 70.3 and 73.6 percent (the nuclear disaster of Fukushima in Japan was likely to have influenced responses for the latter, cf. Ho 2014a).

22   Fu 2020, Question G2, "What do you think is the main source of air pollution in Taiwan?"

23   Map made by the authors with QGIS software, a free and open source Geographic Information System (www.qgis.org/en/site), and complementary data from Natural Earth, a public domain map dataset (www.naturalearthdata.com).

24   As shown in Figures 1–3, Taiwan EPA's website use different colours to symbolize the changing levels of air pollution: from green when safe to yellow, orange, red, violet and purple for levels with risk for public health. A "purple explosion" (*zibao*) has become a common way to raise the alarm on social media. Since 2012, Taiwan EPA's Pollutant Standard Index (PSI) has released measurements of PM10, ozone 3, sulphur and nitrogen dioxide, and carbon monoxide. In 2016, PM2.5 were added to the measurement and the PSI was renamed the Air Quality Index, as in the USA (Jheng 2019, p. 59).

25   Cheng Ling-Jyh, a researcher from Academia Sinica has developed a light measuring device—called Airbox or Location Aware Sensing System (LASS)—which he managed to produce at a low price with the help of Taiwanese electronics makers, for a current network of around 7,000 users (Jheng 2019). On Facebook, another network named *PM2.5 Ziqiuhui: zixun ying gongkai, jujue zang kongqi* ("PM2.5 Self-help Association: Information should be made public, rejecting dirty air") has gathered 24,000 followers as of July 2019. See also Air Clean Taiwan (www.airclean.tw).

26   Taiwan Environmental Information Center, "Zhuang Bing-Jie: Guoguang yun-zhuan hou meinian siwang renshu zengjia 1356 ren" 莊秉潔：國光運轉後 每年死亡人數增加1356人 [Tsuang Ben-Jei: After Guoguang starts operations, the number of deaths will increase by 1356 people per year], 12 November 2010.

27   Our observation of the final court hearing on 30 August 2013 and the written decision of the Taipei District court delivered on 4 September 2013. See also Wild

at Heart Legal Defense Association, "Formosa Plastics v. Prof. Tsuang Ben-Jei Update: Final hearing of the first instance on 29 August," 30 August 2013.

28 Interview with an environmental activist, Taipei, 18 December 2018.

29 We purchased a large amount of online data (a total of 230,512 articles and 8,858,370 comments), for the period from June 1st to December 31, 2018. The data were provided by a platform entitled InfoMiner (launched by LargitData, a private company based in Taipei), which collects information from more than 30,000 sources in Taiwan. The main ones were: 1) news media such as *Liberty Times*, *Apple Daily*, and various Internet news sites (such as *The Reporter*, *The News Lens*, etc.); 2) social media such as Facebook and YouTube; 3) public fora like PTT and Mobile01; and 4) blogs such as Pixnet. In our selection, nearly 40% of articles and comments (93,996) came from Facebook, one third from traditional news outlets (72,937), and 16% from PTT (39,047). The latter is a public forum specific to Taiwan, which dates back to the beginning of the Internet and remains popular in spite of its basic design. Our InfoMiner package included five groups of keywords of our choice (such as "Han Kuo-yu" and a pack of other topics) and the results were distributed by sources, categories and types of emotions; their volume based on the number of publications and comments and their geographic distribution; hot topic analysis and mapping; comparison of keywords; and opinion leaders. Each article could generate up to three or four times more online comments of different sizes (from a few words to long sentences). The top three categories of comments were PTT "gossiping" (1,3 million), PTT "hate politics" (713,798), and remarks on TVBS news, a pro-China television channel (428,763).

References

Batto, Nathan. (2019). "Taiwan's 2018 Election Was About Local Issues, Not China." *The Reporter*, 9 June 2019.

Beck, U. and N. Sznaider. (2006). "Unpacking Cosmopolitanism for the Social Sciences: A Research Agenda." *British Journal of Sociology* 57(1): 1–23.

Callon, M. (1986). "The Sociology of an Actor-Network: The Case of the Electric Vehicle." In *Mapping the Dynamics of Science and Technology*, edited by M. Callon, J. Law, and A. Rip. London: Palgrave Macmillan.

Chan, Chang-Chuan. (2017). "Wo you ziliao zhengming, congtong cuole!" 我有資料證明，總統錯了！[Ms. President, I Can Prove that You Have All Wrong!]. *The Storm Media* 風傳媒, 14 December. www.storm.mg/article/372205.

Chang, Ken-Hui, Fu-Tien Jeng, Ya-Ling Tsai, and Pay-Liam Lin. (2000). "Modeling the Impact of Long-range Transport on Taiwan's Acid Deposition under Different Weather Conditions." *Atmospheric Environment* 34(20): 3281–3295.

Chang, Ying-hwa. (2016). *2010 Taiwan Social Change Survey* (Round 6, Year 1): Environment (C00221_2) [data file]. Available from Survey Research Data Archive, Academia Sinica. doi:10.6141/TW-SRDA-C00221_2-1.

Chen, Chien-Lung, Ben-Jei Tsuang, Rong-Chang Pan, Chia-Ying Tu, Jen-Hui Liu, Pei-Ling Huang, Hsunling Bai, and Man-Ting Cheng. (2002). "Quantification on Source/receptor Relationship of Primary Pollutants and Secondary Aerosols from Ground Sources- Part II. Model Description and Case Study." *Atmospheric Environment* 36: 421–434.

Chen, Jen-Ping, Cheng-En Yang, and I-Chun Tsai. (2015). "Estimation of Foreign Versus Domestic Contributions to Taiwan's Air Pollution." *Atmospheric Environment* 112: 9–19.

Chen, Tu-Fu, Ken-Hui Chang, and Chang-You Tsai. (2014). "Modeling Direct and Indirect Effect of Long Range Transport on Atmospheric PM2.5 Levels." *Atmospheric Environment* 89: 1–9.

Chen, Yi-an and Ming-sho Ho. (2017). "Taiwan fan kongwu yundong 2.0" 臺灣反空污運動 2.0 [The Movement Against Air Pollution 2.0]. In *Neng zenme zhuan: qidong Taiwan nengyuan zhuanxing yaoshi* 能怎麼轉：啟動臺灣能源轉型鑰匙 [How to Turn? Keys for Starting Energy Transition in Taiwan], edited by Chou Kuei-tien and Kuo-hui Chang. Taipei: RSPRC/Liwen.

Chiu, Hsueh-tz'u 邱學慈 (2019). "PTT Source Materials Are All Revealed! How Was 'Han Kuo-yu's Wave' Made?" PTT 原始資料全揭露！「韓流」怎麼造出來的？. *Commonwealth Magazine* 天下雜誌, 9 January 2019 (online).

Chou, Charles C.K., C.-T. Lee, W.-N. Chen, S.-Y. Chang, T.-K. Chen, C.-Y. Lin, and J.-P. Chen. (2007). "Lidar Observations of the Diurnal Variations in the Depth of Urban Mixing Layer: A Case Study on the Air Quality Deterioration in Taipei, Taiwan." *Science of the Total Environment* 374(1): 156–166.

Chou, Charles C.K., C.-Y. Lin, T.-K. Chen, and C.-Y. Young. (2004). "Influence of Long-Range Transport Dust Particles on Local Air Quality: A Case Study on Asian Dust Episodes in Taipei during the Spring of 2002." *Terrestrial Atmospheric and Oceanic Sciences* 15(5): 881–899.

Chou, K. T. (2018). "Introduction: The Cosmopolitan Governance of Energy Transition." In *Energy Transition in East Asia*, edited by Kuei-Tien Chou. London: Routledge, pp. 1–5.

Chou, K.T. and H.M. Liou. (2012). "Cosmopolitan Reform in Science and Technology Governance." *Journal of State and Society* 12: 101–198.

Cui, Yan, Zuo-Feng Zhang, John Froines, Jinkou Zhao, Hua Wang, Shun-Zhang Yu, and Roger Detels. (2003). "Air Pollution and Case Fatality of SARS in the People's Republic of China: An Ecologic Study." *Environmental Health: A Global Access Science Source* 2: 15.

Edwards, Paul N. (2010). *A Vast Machine: Computer Models, Climate Data, and the Politics of Global Warming.* Cambridge, MA: MIT Press.

Fell, Dafydd J. (2016). "The China Impact on Taiwan's Elections: Cross-strait Economic Integration Through the Lens of Election Advertising." In *Taiwan and the 'China Impact': Challenges and Opportunities*, edited by G. Schubert. London: Routledge, pp. 53–69.

Friedman, Lisa. 2020. "New Research Links Air Pollution to Higher Coronavirus Death Rates." *New York Times*, 7 April.

Fu, Yang-Chih. (2020). *2019 Taiwan Social Change Survey (Round 7, Year 5): Technology and Risk Society* (Restricted Access Data) (R090063) [data file]. Available from the Survey Research Data Archive, Academia Sinica. doi:10.6141/TW-SRDA-R090063-1.

Grano, Simona. (2015). *Environmental Governance in Taiwan: A New Generation of Activists and Stakeholders.* London: Routledge.

Ho, Ming-sho. (2005). "Protest as Community Revival: Folk Religion in a Taiwanese Anti-Pollution Movement." *African and Asian Studies* 4(3): 237–269.

Ho, Ming-sho. (2006). *Lüse minzhu: Taiwan huanjing yundong de yanjiu* 綠色民主：台灣環境運動的研究 [Green Democracy: A Study on Taiwan's Environmental Movement]. Taipei: Socio Publishing.

Ho, Ming-sho. (2014a). "The Fukushima Effect: Explaining the Resurgence of the Anti-nuclear Movement in Taiwan." *Environmental Politics* 23(6): 965–983.

Ho, Ming-sho. (2014b). "Resisting Naphtha Crackers: A Historical Survey of Environmental Politics in Taiwan." *China Perspectives*: 5–14.

IQAir. (2020). *2019 World Air Quality Report: Region & City PM2.5 Ranking.* www.iqair.com

Jheng, Gordon Shihao. (2019). "Huiji gongming liliang de kongwu zhandouqi: kongqi hezi" 匯集公民力量的空污戰鬥器：空氣盒子 [Air Pollution Fighting Machines that Bring Citizens Together: Airboxes]. In *Richang shenghuo de nengyuan geming* 日常生活的能源革命 [Energy Revolution in Daily Life], edited by Chou Kuei-tien et al. Taipei: Springhill & RSPRC, pp. 76–93.

Jobin, P. (2021). "Our 'Good Neighbor' Formosa Plastics: Petrochemical Damage(s) and the Meanings of Money." *Environmental Sociology* 7(1): 40–53.

Latour, Bruno. (1987). *Science in Action: How to Follow Scientists and Engineers through Society.* Cambridge, MA: Harvard University Press.

Latour, Bruno. (2017a). *Down to Earth: Politics in the New Climatic Regime.* London: Polity.

Latour, Bruno. (2017b). *Facing Gaïa. Eight Lectures on the New Climatic Regime.* Cambridge: Polity.

Law, John and John Hassard, eds. (1999). *Actor Network and After.* Oxford: Blackwell.

Lee, Taedong and Wooyeal Paik. (2020). "Asymmetric Barriers in Atmospheric Politics of Transboundary Air Pollution: A Case of Particulate Matter (PM) Cooperation Between China and South Korea." *International Environment Agreements* 20: 123–140.

Lidskog, Rolf and Göran Sundqvist, eds. (2011). *Governing the Air: The Dynamics of Science, Policy, and Citizen Interaction.* Cambridge, MA: MIT Press.

Lin, Chuan-Yao, Shaw C. Liua, Charles C.-K. Chou, Saint-Jer Huang, Chung-Ming Liu, Ching-Huei Kuo, and Chea-Yuan Young. (2005). "Long-range Transport of Aerosols and Their Impact on the Air Quality of Taiwan." *Atmospheric Environment* 39: 6066–6076.

Liu, Yiting. (2019). "Ziji de kongqi ziji jiu: Jiayi yu Pulizheng fankongwu gushi" 自己的空氣自己救：嘉義市與埔里鎮反空污故事 [Solving Our Air Problem by Ourselves: The Anti-air Pollution Stories of Chiayi City and Puli Township]. In *Richang shenghuo de nengyuan geming* 日常生活的能源革命 [Energy Revolution in Daily Life], edited by Chou Kuei-tien et al. Taipei: Springhill & RSPRC, pp. 53–75.

Lu, Hsin-yi. (2016). "Tudi, shequn, xinyang: jiexi sumin huanjing lunshu" 土地、社群、信仰：解析俗民環境論述 [Land, Community, and Faith: An Analytical Model of Folk Environmental Discourse]. *Taiwanese Journal for Studies of Science, Technology and Medicine* 科技、醫療與社會 22: 6–108.

Lu, Hung-Yi, John Kennedy Mwangi, Lin-Chi Wang, Yee-Lin Wu, Chong-Yu Tseng, and Ken-Hui Chang. (2016). "Atmospheric PM2.5 Characteristics and Long-Term Trends in Tainan City, Southern Taiwan." *Aerosol and Air Quality Research* 16: 2488–2511.

Middleton, Carl and Jeremy Allouche. (2016). "Watershed or Powershed? Critical Hydropolitics, China and the 'Lancang-Mekong Cooperation Framework'." *The International Spectator* 51(3): 100–117.

Moore, Scott. (2018). "China's Domestic Hydropolitics: An Assessment and Implications for International Transboundary Dynamics." *International Journal of Water Resources Development* 34(5): 732–746.

Schubert, Gunter, ed. (2016). *Taiwan and the 'China Impact': Challenges and Opportunities.* London: Routledge.

Shie, R. H., and C. C. Chan. (2013). "Tracking Hazardous Air Pollutants from a Refinery Fire by Applying On-line and Off-line Air Monitoring and Back Trajectory Modeling." *Journal of Hazardous Materials* 261: 72–82.

Tsuang, Ben-Jei. (2018). "Laiyuanzhang nin cuole, ranmei dianchang juedui shi zhong wuran!" 賴院長您錯了，燃煤電廠絕對是重汙染！ [President Lai, You Are Wrong. Coal-fired Power Plants Are Definitely Heavy Polluters!]. *Common Wealth Magazine* 天下雜誌, 16 October 2018. opinion.cw.com.tw/blog/profile/52/article/7374.

Tu, Wenling. (2019). "Combatting Air Pollution Through Data Generation and Reinterpretation: Community Air Monitoring in Taiwan." *East Asian Science, Technology and Society: An International Journal* 13(2): 235–255.

Wu, Jieh-min. (2020). "Taiwan's Election is a Vote About China." *The New York Times*, 10 January.

Wu, Jieh-min and Mei Liao. (2015). "From Unification-Independence Divide to the China Factor: How Changing Political Identity Influences Voting Behavior." *Taiwanese Sociology* 29: 89–132 (In Chinese).

Wu, Nai-Teh. (2016). "Political Competition Framed by the China Factor? Looking Beyond the 2012 Presidential Election." In *Taiwan and the 'China Impact': Challenges and Opportunities,* edited by G.Schubert. London: Routledge, pp. 130–148.

Yang, Wen-Shan. (2021). *2017 Social Image Survey I (C00330)* [data file]. Available from Survey Research Data Archive, Academia Sinica. doi:10.6141/TW-SRDA-C00330-1.

Yin, Yu-huan. (2017). "Kongwu dou shi Zhongguo haide? Huanbaoshu weiwai yanjiu dalian: Taiwan paifang liu cheng liu wuran, Zhongguo zhan san cheng" 空汙都是中國害的？環保署委外研究打臉：台灣排放6成6汙染、中國佔3成 [Is Air Pollution Caused by China? EPA's Own Outsourced Research Slaps its Face: Taiwan Emits 60% of Its Pollution, China Accounts for 30%]. *The Storm Media* 風傳媒, 14 December. www.storm.mg/article/372438.

Yuan, T.-H., Y.-C. Shen, R.-H. Shie, S.-H. Hung, C.-F. Chen, and C.-C. Chan. (2018). "Increased Cancers among Residents Living in the Neighborhood of a Petrochemical Complex: A 12-year Retrospective Cohort Study." *International Journal of Hygiene and Environmental Health* 221(2): 308–314.

5 Asymmetric barriers in atmospheric politics of transboundary air pollution

A case of particulate matter (PM) cooperation between China and South Korea

Taedong Lee and Wooyeal Paik

Acknowledgment: This research has been supported financially by the Posco TJ Park Foundation's Research Grants for Asian Studies and Korean National Research Foundation (2019S1A5A2A01047251). We revised the manuscript that was published at International Environmental Agreement 20: 123–140 (2020).

5.1 Introduction

On March 7, 2019, the Korean government took emergency measures to reduce particulate matters (PM) 2.5 which rose to a bad or very bad level (81–150 ug/m^3 or 151 ug/m^3) for seven days in a row.[1] Most South Koreans blame China for being the primary source of PM2.5, PM10, and yellow dust emissions. Satellite images of PM in the media vividly show the high PM density as well as the direction of PM-laden wind moving from China to the Korean peninsula. Scientific research has also pointed to the "strong influence of the east-central Chinese emission on South Korea, at maximum 200% over the pathway of long-range transport of aerosol pollution compared to the mean condition" (Lee et al. 2019), particularly in "multi days severe air pollution in cold seasons" (Oh et al. 2015). The citizen-led petition to the Blue House—the Presidential Office of South Korea—on this issue attracted 278,128 signatures in just one month, from March 24 to April 23, 2018 (Presidential Office 2018), forcing the Presidential Office to engage with China diplomatically.

While severe transboundary air pollution in Northeast Asia demands urgent regional cooperation (Kim 2007; Shim 2017; Yarime and Li 2018), both bilateral and multilateral cooperation still remains rudimentary. Despite the series of meetings and initiatives including the Acid Deposition Monitoring Network in East Asia (EANET) and Long-range Transboundary Air (LTP), China and Korea have not reached a consensus even on the scientific facts and related responsibility, to say nothing of a binding agreement (Kim and Kim 2018; Shim 2017; Yoon 2007). Compared to cooperation on transboundary air pollution in other regions such as

DOI: 10.4324/9781003211747-7

Europe (Fraenkel 1989; Tuinstra et al. 2006) and Southeast Asia (Nurhidayah et al. 2015), there has been little bilateral cooperation. Why is there so little consensus on the causes and effects of transboundary air pollution in Northeast Asia, particularly between China and South Korea? Furthermore, why is there neither a bilateral nor a multilateral agreement on transboundary air pollution in Northeast Asia?[2]

Although existing studies have argued that regional cooperation is urgently needed to tackle transboundary air pollution (Kim 2007; Min 2001; Shim 2017; Yarime and Li 2018) along with other issues (Haas 1990), few studies have operationalized atmospheric cooperation steps and theorized the underlying logic of asymmetry as a barrier to further cooperation. The existing literature has not systematically analyzed the steps of transboundary air pollution cooperation. Given that air quality degradation and its detrimental impacts on neighboring countries are worsening around the world, it is imperative to identify an adequate framework with which to analyze the degree of bilateral and multilateral transboundary cooperation. This study aims to explain the barriers around transboundary air pollution cooperation in general and to test the explanation empirically through a case study of China and South Korea. Our conceptualization of the process of cooperation in atmospheric politics and our illustration of asymmetric barriers—in state capacity, economic interests, domestic pressure, and international pressure—theoretically and empirically contribute to the field of global environmental politics and China and Korea studies. We theoretically find that asymmetries (as independent variables) are likely to impede the steps of bilateral environmental cooperation (as a dependent variable). Empirically, this study contributes to the understanding of challenges in transboundary cooperation between China and South Korea with a case study. In echoing the key concepts of this book, our research explores interactions between China factors and neighboring country to respond to transboundary risks. In the region, air is common. Cosmopolitan governance and cooperation are critical to overcoming scientific uncertainty and geopolitical barriers. This systematic analysis points to policy suggestions that improve the chances of reducing transboundary air pollution through regional cooperation

To this end, Section 5.2 analyzes the nature of transboundary atmospheric politics and operationalizes cooperation steps as the dependent variable. We also theorize asymmetric barriers in state power, economic interests, and domestic pressures as well as a lack of international pressure as the independent variable to explain the lack of cooperation between the two countries. Section 5.3 empirically analyzes the suggested causal relation and its consequences. Section 5.4 concludes the paper and provides policy suggestions to improve transboundary air pollution cooperation between the two countries, as well as between countries across the world.

5.2 Atmospheric political cooperation and its barriers

5.2.1 *Cooperation steps for atmospheric politics*

No one can stop breathing for more than a few minutes. Some may take clean air for granted, but in many parts of the world, this is not possible. Thus, ensuring high air quality has been a core task of environmental management. However, the atmosphere cannot be contained by human-made boundaries. Dust, particulate matter, and acid rain containing pollutants easily cross state borders. Transboundary air pollution demands international cooperation, yet bilateral or multilateral cooperation on air pollution is a challenge. In particular, cooperation between countries with asymmetries in power, interests, and domestic and international pressure faces many difficulties.

To explain this phenomenon, we begin by operationalizing cooperation steps and theorizing asymmetric barriers in cooperation on atmospheric politics, focusing on the transboundary air pollution issue. With the well-developed literature on hydropolitics as a guide (Daoudy 2009; Dinar 2009; Han 2017; Warner and Zawahri 2012), we identify the components of atmospheric politics, examining the degree of cooperation and contention among stakeholders in a shared atmosphere with the following questions: (1) What are the characteristics of atmospheric politics? (2) What are the frameworks by which to analyze countries' cooperation on atmospheric politics? (3) How do these characteristics and frameworks form different dynamics that revolve around the countries' (or political entities') cooperation on agreeing to decrease transboundary air pollution?

We define atmospheric politics as the authoritative allocation of value with respect to the atmosphere. The scope of atmospheric politics covers global issues (such as climate change and ozone layer depletion), regional issues (such as transboundary air pollution and acid rain), and local issues (such as air quality degradation) (Mitchell 2010). This study focuses on regional atmospheric politics, particularly transboundary air pollution. Atmospheric politics necessarily involves problems with amorphous attributes (blurred boundaries and multiple causes), issues of quality rather than quantity (fewer incentives for economic cooperation), and direct links to economic development based on energy use.

First, the amorphous aspects of atmospheric politics tend to lead to a mismatch of jurisdictions. While sovereign airspace exists, atmospheric boundaries are blurred, and transboundary air flows are multi-directional. This can create difficulties in establishing clear jurisdiction, reasonability, causes and costs of damages, and coordination between science and policy (Tuinstra et al. 2006). Second, the quality is more important than the quantity of transboundary air issues. Without proper schemes, air pollution quantity is not financially beneficial for trade. In contrast, in hydropolitics, water quantity can be a clue for cooperation (as well as conflict) through sharing water resources (irrigation, industrial use, or hydropower generation). Cooperation with a large volume of water resources for electricity generation or other water usage can facilitate mutual and direct economic benefits for participating countries. However, unlike hydropolitics, atmospheric politics does not generate immediate and monetary incentives for cooperation because states do not gain anything material or visible. Of course, cooperation in atmospheric politics could beget benefits such as public health and eco-system enhancement, but these incentives and outputs tend to be long term and are not easily convertible to monetary value. Third, emissions of air pollutants (and bad air quality) are closely linked to economic and industrial development. When more economic development (including increases in energy consumption and motor vehicles) is pursued, greater emissions are produced (Harris and Lee 2017). To reduce air pollution, substantial regulatory and monetary efforts should be in place with proper linkages between governance, institutions, and actors (Selin and VanDeveer 2003). In summary, atmospheric politics, including regional transboundary air pollution, global climate change, and ozone depletion, concerns problems that are challenging to solve collectively.

The process of environmental cooperation has been discussed in the literature. Hass (1990), for example, proposed the three phases of policy activities; agenda setting, international policies; and national policy response. Agenda setting is to identify problems for collective response. International policies aim to coordinate policy measures for collective application. National policy response is to comply with international measures. Mitchell (2010) added some other components in international environmental negotiation. To set the agenda, it needs to build up knowledge, concern, and urgency. In the process of international policymaking, mutually acceptable goals and policies should be generated. After maintaining and promoting momentum, a regulatory or procedural form of institution can be initiated and strengthened (O'Neill 2009).

Based on the phase of environmental cooperation (Haas 1990; Mitchell 2010; O'Neill 2009; Yarime and Li 2018), we operationalize the process of environmental cooperation on transboundary atmospheric politics in four steps. We develop this process to analyze transboundary air pollution cooperation at bilateral and multilateral levels, but this can be utilized as a step in identifying general environmental cooperation.

The first stage is issue identification, in which bilateral or multilateral discussions about atmospheric politics and the causes and effects of transboundary air pollution begin (Mitchell 2010; Yarime and Li 2018). Paying attention to shared atmospheric problems and initiating discussion is the first stage in cooperation in atmospheric politics. In this stage, it is critical to identify the parties' shared issues. Identifying stakeholders is also imperative to beginning discussions.

The second stage involves collaborative research, shared recognition of air pollution responsibility, and a search for policy options. This stage aims to build scientific, political, and economic consensus on the issues identified (Kim 2007). This step can be facilitated by collaborative research to identify the causes and effects of transboundary air pollution. Collaborative research consists of

Figure 5.1 Cooperation steps for atmospheric politics of transboundary air pollution

(1) monitoring emissions data (sources and amount) and air and precipitation quality, and (2) modeling air pollutant flows and the causes and effects of pollutants (Kim 2014, 156) based on scientific data and research (Lee et al. 2019). It is critical to identify where, how much, and which air pollutants come from and flow to. In addition to natural scientific research, social science research to examine policy options and economic viability is also required (Mitchell 2010). All collaborative research outputs should be shared and made publicly available in order to proceed toward a binding agreement. Otherwise, uncertainty in scientific findings in natural and social science may impede progress. Related to knowledge and information, a number of studies also mention the absence of an epistemic community as a factor that explains the poor level of environmental cooperation in Asia as compared to that in other regions (Kim 2007).

After setting the agenda in the first and second steps, the third step is to make a legally binding agreement on air pollution management through negotiation process (Mitchell 2010; O'Neill 2009). This multilateral or bilateral environmental agreement should set out rights and obligations that include procedural details, implementation schemes, organizational cooperation (institutional) setting, funding, and reporting requirements (Min 2001). Policy coordination lays out the procedure for environmental cooperation by setting out which state parties have what kind of responsibilities.

The final step is to implement an agreement on atmospheric politics. With funding and institutional schemes, the concerned parties translate the agreement into international and domestic actions. Stakeholders implement the enacted policy options with organization and resources. This stage includes conducting domestic and international measures to mitigate transboundary air pollution, monitoring the performance of international cooperation and domestic action, and making revisions to address unexpected drawbacks.

Proposed cooperation steps for atmospheric politics of transboundary air pollution can be applied to analyze past efforts including the USA-Canada Air Quality Agreement (Roelofs 1993) and Finland-USSR. This framework can be also useful to analyze and propose cooperation schemes to address current bilateral transboundary air pollution among Pakistan-India-Bangladesh, other countries and regions.

5.2.2   *Asymmetric barriers to advancing cooperation in transboundary air pollution*

Three asymmetries—in state capacity, economic interests, and domestic pressure—and a lack of international pressure are key factors in producing conflict and cooperation in atmospheric politics, which revolves around transboundary air pollution between "source" and "receptor" countries.

First, asymmetry in general state power stems from differences in state power between the involved countries. State power depends on multiple factors, such as economy (e.g., Gross Domestic Product), military, territory, population, and cultural heritage. This asymmetry allows a hegemonic state located in the

"upstream" or source position to evade or prevent advancing four-step cooperation with countries in the "downstream" or affected position (Warner and
Zawahri 2012). If their state powers significantly differ, the weaker state can
neither force nor financially help the stronger state to agree and implement an
agreement. As is witnessed in hydropolitical dynamics, following the simple logic
of realism in international relations (Hass 1990; O'Neill 2009), which analyzes
the conflictual process among multiple states with general power asymmetries
and ensuing consequences, an upstream hegemon has more incentive to ignore
the afflicted countries with lesser state capacity and less incentive to cooperate
with them (Dinar 2009). In a similar vein but from a different angle, this
asymmetry of state capacity produces a "politics of over-attention and inatten-
tion" (Shin et al. 2016; Womack 2015). This is a situation in which weaker
countries give most of their attention to bilateral atmospheric political interaction
while trying to force their larger counterparts to cooperate. Stronger countries,
however, are occupied with other international security and economic issues
and thus do not devote as much attention to the conflictual relationship (Shin
et al. 2016; Stephen 2009; Womack 2015). We hypothesize that a higher level
of asymmetry in state capacity likely hinders progress in transboundary air pollution cooperation in Northeast Asia.

The second asymmetry in economic interests is another factor that influences
the phased development of cooperation in transboundary air pollution atmospheric
politics (Fraenkel 1989). If countries in the region shared similar costs and benefits,
as in the hydropolitical cases, cooperation is likely. Despite state capacity differences
between stronger hegemons upstream and weaker neighbors downstream, environmental regimes for cooperation occasionally work through the co-benefits of
joint action to create and share added value according to a neoliberal theory
perspective, which emphasizes that economic co-benefits propel multiple states
to cooperate on a certain bilateral and/or multilateral issue (Dinar 2009; Han
2017; Hussein and Mattia 2017; Kim 2014). However, asymmetries in economic
costs and benefits based on the polluter-pays principle hinder cooperation (Perrin
and Bernauer 2010). We assume that, unlike hydropolitical cooperation, which
produces electricity, water irrigation, environmental diversity, and other shared
economic benefits, in atmospheric politics, transboundary air, whether clean or
dirty, creates few if any bilateral economic opportunities or international markets.
This means that each country—be it a source, a receptor, or both—is more likely
to obtain economic benefits if the source country reduces domestic air pollution.
However, any type of added value such as electricity in hydropolitics cannot be
created in atmospheric politics, aside from potential benefits from reducing environmental risks. Therefore, one of the two dimensions of economic co-interests
from hydropolitical cooperation—not added value but reduced pollution—is the
only dimension for economic interests that typically matters more in a receptor
country. We argue that a higher level of asymmetry in economic interests is likely
to hamper the advancement of transboundary air pollution cooperation.[3]

The third factor, that of asymmetry in domestic pressure, is based upon the
established literature on the interaction between domestic politics and

international politics. Domestic and international politics frequently, if not always, interact (Gourevitch 2002; Putnam 1988). Domestic changes prompt changes in the international context and vice versa; the causal directions usually run both ways. A set of domestic variations in the countries directly involved play an important role in the foreign policy decision-making process. This asymmetry is fundamental for state leaders to take transboundary air pollution seriously and try to solve the issue in an international framework. The source country's domestic audience cares about the serious air pollution within its territorial boundary because it directly harms their quality of life, and only rarely because it also harms the quality of life in neighboring countries. They do not care about and so ignore the effect of transboundary air pollution in foreign lands. By contrast, the receptor country's domestic audience puts enormous political pressure on their government to bilaterally negotiate with the source country to reduce its domestic air pollution or multilaterally attract international attention and cooperation. We argue that asymmetries in domestic pressure at the giving and receiving ends are likely to impede international cooperation on transboundary air pollution.[4]

In addition to these three asymmetries, the last element in atmospheric political cooperation on transboundary air pollution is international pressure. Neoliberal scholars of international relations pay great attention to the role of international organizations or regional integration in mediating between stakeholders in international conflicts (Keohane and Nye 2012; Luterbacher and Sprinz 2001). There are many examples of global—often regional—organizations that mediate and settle serious conflicts of interest among the member countries of a region. Moreover, such a binding international regime puts substantial pressure on larger and stronger countries, often called regional hegemons, to protect smaller neighboring countries through either international legal mechanisms or the collective efforts of smaller countries in the region. If there is a binding international organization or integration structure such as the European Union (EU) or the Association of Southeast Asian Nations (ASEAN) in the region of the transboundary air pollution conflict, phased cooperation on reducing transboundary air pollution is more likely to happen.

5.3   Transboundary air pollution cooperation between China and South Korea

This study uses a primarily qualitative methodology because it aims to understand the causes and effects of the degree of phased transboundary air pollution cooperation among countries with an in-depth case study. This study depends on qualitative data—(1) archival data (government documents, statistical data, policy briefs, media reports, etc.) and (2) interview data collection and analysis (utilizing key interviews with sixteen experts, officials, and journalists among many others). Interviewees were selected due to their expertise in environmental cooperation and transnational air pollution issues (published journal articles, news articles for scholars and journalists, and primary tasks for public officials).

We conducted semi-structured interviews with open-ended questions. The data were collected from the targeted countries—South Korea (Seoul) and China (Nanjing, Beijing, and Jilin Province)—a various (national, local, and partly individual) levels during recurring fieldwork from June 2018 to August 2019.[5]

5.3.1 Transboundary air pollution cooperation between South Korea and China

Based on the theoretical framework, we empirically examine the process of transboundary air pollution cooperation between South Korea and China. This study focuses on the barriers to cooperation for agreement on the causes and effects of transboundary air pollution between South Korea and China rather than real actions that have been taken to deal with such pollution, as no actions have yet been taken. In the above analytical scheme of cooperation steps for atmospheric politics, we find that South Korea and China have achieved some attributes of the intermediate stage of collaborative research. South Korea and China bilaterally have not achieved a legally binding agreement (the third step) or implementation (the fourth step). The concrete contents are as follows.

Bilateral environment cooperation between the government of the Republic of Korea and the government of the People's Republic of China began in 1993 with the signing of the Agreement on Environmental Cooperation. After the collapse of the Soviet Union and expanded diplomatic relations with former communist countries, South Korea regarded environmental cooperation with China as fundamental for extending and deepening the countries' bilateral relationship since the impacts of transboundary pollution are direct. In the agreement, both governments share broad and somewhat symbolic issue identification, such as "coping with global environmental degradation threatening human survival" (MoE 1993).

Ten years later, in 2003, the agreement identified the primary ministries for cooperation with a Memorandum of Understanding (MOU) between the Ministry of Environment of the Republic of Korea and the State Environmental Protection Administration of the People's Republic of China on Environmental Cooperation. The symbolic agreement required identifying the responsible party for environmental cooperation on a range of issues. The MOU between the Ministry of Environment (South Korea) and the State Environmental Protection Administration (China) aimed to solidify the 1993 agreement in order to recognize increasing threats from transboundary air pollution (yellow dust) and desertification. The MOU also stated that both ministries recognized the benefits of collaborative research on long-range transboundary air pollution (MoE 2003). However, there was no concrete plan or schedule for collaborative research.

In 2005, more focused cooperation for transboundary air pollution was discussed under an arrangement between the Ministry of the Environment of the Republic of Korea and the State Environmental Protection Administration of the People's Republic of China on Ground Monitoring and Information Exchange for Dust and Sand Storms (MoE 2005). This arrangement proposed a monitoring and alarm system to track transboundary air pollution. However, the results of the collaborative research and information/data exchange were not made public. After the arrangement came a series of MOUs for collaborative research and potential agreements to tackle transboundary air pollution,[6] but these efforts were in vain. Kim and Kim (2018) identified a lack of actual collaboration between China and South Korea on transboundary air pollution, despite a number of MOUs. Our interviews with South Korean experts and public officials confirmed a lack of collaborative research and sharing of research outputs between South Korea and China.[7] Furthermore, scientific studies on transboundary air pollution from Chinese experts have mostly focused on China's domestic air pollution rather than regional (Asian or international) pollution (for example, Gu and Yim 2016).

Instead, the Ministry of Environment of South Korea collaborated with the US space agency NASA (National Aeronautics and Space Administration). Their joint project, KORUS-AQ (South Korea-US Domestic Air Quality Collaborative Research), announced that approximately 52 percent of PM2.5 came from domestic sources; 34 percent from China; 9 percent from North Korea; and 6 percent from other places in 2017 (NIER 2017). This scientific research project was conducted through international cooperation, but without China's participation.[8]

A number of proposals for collaborative research between South Korea and China have been announced. For instance, the Korean National Institution of Environmental Research and the Chinese Research Agency for Environmental Science (CRAES) signed the MOU for the establishment of a Collaborative Research Team between South Korea and China in 2015. However, in contrast with the public announcement of collaborative research outcomes between South Korea and the USA, China refused to publicly announce the outcomes of collaborative research on China, Japan, and South Korea's transboundary air pollution (LTP: Long-range Transboundary Air Pollutants in Northeast Asia).[9] While collaborative research has been conducted following suggestions from the Tripartite Environmental Ministers' Meeting among South Korea, China, and Japan (TEMM) and other MOUs, China argued that collaborative research results are "expected" to be published rather than "agreed" in the MOU. China also insisted that the data for research were obsolete (data from China from 2008 and 2010, compared to data from South Korea from 2013) and therefore too unreliable to present publicly. Thus, South Korea and China failed to publicly share collaborative research that could identify the causes, trajectories, and effects of transboundary air pollution in East Asia.[10] The spokesperson of China's Ministry of Ecology and Environment recently argued that PM in Seoul comes from Seoul itself rather than China. He insisted that the PM level in Chinese cities has dramatically improved while the level in Seoul has worsened. In summary, Chinese officials publicly announced that there was little evidence that Seoul's PM was caused by China (Chinese Ministry of Foreign Affairs 2017; Kim and Kim 2018). The Minister of Environment of Korea rejected the Chinese argument, saying that data have shown that high levels of PM in South Korea were coming from China and other foreign sources (Cho 2019).

In applying the analytic framework of cooperation steps/phases for atmospheric politics, South Korea and China have managed to identify transboundary air pollution as a shared environmental problem. From the beginning of the bilateral agreement, transboundary air pollution has been on the list of issues requiring cooperation. Both countries are trying to find multiple stakeholders, encompassing ministries, businesses and industries, and researchers. At most environmental meetings, transboundary yellow dust and PM issues have been discussed as key agenda items.

However, China and South Korea have not reached the second step for collaborative research. Based on the shared recognition of the importance of collaborative research, it is critical that the two countries conduct and share natural and social scientific analysis. While substantial emphasis has been placed on collaborative research, China and South Korea have not presented research results stating the cause, trajectories, and effects of transboundary air pollutants. Without common and shared beliefs regarding the causal mechanisms of air pollution, it will be challenging to propose policy options as well as a binding agreement with mandates.

5.3.2 Hindering asymmetries in cooperation on transboundary air pollution between China and South Korea

We empirically apply the proposed theory of asymmetry to a case of transboundary air pollution cooperation between South Korea and China. As the previous section illustrated, this study focuses on the lower stage of bilateral cooperation in the phased development of transboundary air pollution settlement: the countries in question have passed the first stage of issue identification and are stuck at the second stage of collaborative research regarding the causes and effects of transboundary air pollution. Thus, this empirical study does not deal with the third and fourth stages, as they have not yet occurred.

The failure to establish cooperation to settle transboundary air pollution between South Korea and China stems from the following factors. The suggested three asymmetric categories—state power, economic interests, and domestic pressure—and the element of international pressure play a critical role in producing such a result. On the one hand, China's much larger state capacity, the lack of common economic interests related to the issue, the lack of domestic pressure in China, and little international pressure have hindered the two countries from settling the issue of transboundary air pollution of PM10 and 2.5 from China to South Korea throughout the 2010s. On the other hand, the mounting domestic pressure in South Korea to settle this issue has pushed the South Korean government to accelerate the phased development of settlement (Kim 2018; Oh 2019). The aforementioned scientific research shows that this air pollution was a serious problem in South Korea well before the 2010s. However, the severity of the pollution and renewed public awareness have forced the

South Korean government to redouble its efforts to complain to and negotiate with China since 2016, as demonstrated by the aforementioned petition to the Presidential Office. These asymmetric factors have been combined with a lack of bilateral cooperation and deepened the two countries' environmental conflict.

First, there is an obvious asymmetry in the state power of China and South Korea. China's territory is 100 times the size of South Korea; its population is twenty-eight times as large as that of South Korea. Differences in their economic and military state capacities are substantial. The GDP ratio is almost ten to one—China's was USD12.2 trillion and South Korea's was USD1.5 trillion in 2017. The military spending ratio is eight to one—China's was USD225.7 billion and South Korea's was USD37.3 billion, not counting nuclear weapons and aircraft carriers, in 2017.[11] Moreover, the South Korean economy has become increasingly overdependent on China since the 2010s, creating another economic asymmetry; the mutual trade ratio is 30 percent for South Korea and 8 percent for China.[12] In other words, South Korea is significantly less powerful in military capacity than China, impeding it from coercing China to cooperate. Even though China's military superiority is not directly used for environmental gains in this case, it allows China not to seriously consider more cooperation with South Korea on settling the environmental conflicts. It also has significantly fewer economic resources than China, meaning that it cannot subsidize or finance China to reduce its air pollution or to cooperate in the phased development of the PM issue settlement.

As the theory of asymmetry in international relations predicts (Shin et al. 2016; Womack 2015), a larger country, often referred to as a "regional hegemon," gives little or no attention to the conflictual issues of transboundary air pollution with its smaller neighbor, which usually generates excessive attention on the issue among the elite and the public. In this context, even though the domestically pressured South Korean government demands that the Chinese government cooperates on the PM issue, thus far its demands have been in vain (Huanqiushibao 2017; Kim 2018; President Office 2018).[13] China has not been very proactive in the settlement of the PM issue with South Korea (Huanqiushibao 2017);[14] for the Chinese government, this transboundary air pollution issue is minor in comparison with other issues it faces, such as North Korean nuclearization, the trade war against the USA, and frictions involving the Belt and Road Initiative.

Another dimension of asymmetry is the lack of economic opportunities to create added value and environmental advantages that are mutually beneficial for South Korea and China. As we discussed above, the settlement of pollution in hydropolitics can generate substantial economic added value in addition to the general ecological and hygienic benefits for both source and receptor countries. However, in the case of transboundary PM pollution between South Korea and China, there are few if any mutually beneficial economic interests.[15] If the PM transfer from China to South Korea is dramatically reduced, South Korea will receive enormous economic benefits through improved public health. Last few years, the South Korean national and local government has increased its budget to improve the deteriorating quality of air, mainly due to PM2.5 and 10. The government increased the budget to cut the PM pollution in South Korea from 2.3 trillion Korean Won (20 billion USD) in 2019 to 4 trillion Korean Won (3.7 billion USD) in 2020, 74.6% increase in just one year (Park 2019). The financial cost is significant for South Korea, but it does not directly benefit China.

Moreover, the fact that atmospheric politics lack an economic mechanism to generate cooperation (unlike hydropolitics, which impacts fishery resources, electricity, etc.) exacerbates asymmetry between the countries, thus hindering international cooperation. One economic incentive for China to reduce transboundary air pollution might be targeted financial and technological support from South Korea, as South Korea attempted to provide in 2016 and 2017.[16] However, these efforts had negligible effects in the Chinese cities in which they were tested because the amount of support was insufficient. As such, South Korea cannot create the economic incentive structure for China to reduce the transboundary PM so far.

Third, there is an obvious asymmetry in the pressure of the countries' domestic audiences to settle this PM transfer issue—it is high in South Korea but almost nonexistent in China. South Korea is the receptor, with a consolidating democracy, developed economy, and scientific knowledge on transboundary air pollution among the domestic audience.[17] The Chinese public, however, stands in the opposite position and criticizes its South Korean counterparts on this issue. The public, led by the national media (Dong 2018),[18] deny any transboundary PM from their country to South Korea and call South Korea's claims groundless (Dong 2017; Huanqiushibao 2017; Liu and Wang 2017), even though these Chinese citizens heavily criticize the identical air pollution in their domestic environment and have placed heavy pressure on their central and local government to reduce it. Instead, they point out that South Korea itself generates a substantial amount of PM, according to Chinese, South Korean, and other investigations. Above all, most citizens, including Chinese experts who study the Korean peninsula and who are most knowledgeable about bilateral relations, neither recognize nor care about the issue.[19] Kim and Kim (2018)'s content analysis from 1996 to 2016 found that news reports and scientific research from China have not linked air pollution in China to its transboundary impacts on Northeast Asia.

Another element that hinders cooperation on this issue is the lack of international regime pressure on China. Northeast Asia has no reliable or functioning regionally integrating organization such as the EU in Europe (Fraenkel 1989; Tuinstra et al. 2006) or, to a lesser extent, ASEAN in Southeast Asia (Nurhidayah et al. 2015). This type of international organization among countries with less asymmetrical state power usually possesses binding power to help member countries negotiate and settle conflictual environmental issues. Transboundary air pollution is a good example of issue-specific buffering to accept the guidelines for dispute resolution or to submit a dispute to an international tribunal. If the

PM issue were framed in a broader multinational or international regional context, the bilateral asymmetry would be reduced, and the phased development of the PM issue settlement could be accelerated. Nevertheless, South Korea's ongoing attempts to work with Japan to encourage cooperation from China to investigate this particular atmospheric political issue in the 2010s via the Tripartite Environmental Ministers Meeting has not had substantial success. Moreover, a more comprehensive multilateral regional organization, the North-East Asian Subregional Program for Environmental Cooperation (NEASPEC), does not yet have an effective regional architecture with binding institutions or treaties in Northeast Asia.

5.3.3 *Discussion and policy suggestions*

Given the barriers, in the foreseeable future, cooperation through bilateral or multilateral international negotiations to settle transboundary air pollution between these two countries hardly move beyond the second step of collaborative research to produce shared recognition of the causes and effects and then consider policy options, not to mention the third step of binding agreement and implementation in the foreseeable future.[20]

Instead, it appears that the solution to such a conundrum should come from both the international dimension and domestic Chinese domestic politics. Note that there is a "positive externality" factor of the source-transferor country's own efforts to reduce air pollution in its territory. In this context, a more plausible solution for transboundary air pollution might be motivated not by international negotiations and cooperation but by the pollution transferor's domestic needs to reduce it for its own sake. In addition, domestic pressure to reduce regional as well as intranational transboundary air pollution across one's own territory—for example, from province A to province B—may motivate the central and provincial governments to take actions to curb the pollution and may also be effective in the international arena. The positive externality of these domestic conflicts would be another critical factor in the process of international transboundary air pollution cooperation.

In making progress in bilateral cooperation with South Korea to reduce the transfer of polluted air, the already mounting domestic pressure in China and its government's proactive and serious responses to reduce PM would be a key under the international stalemate (Ahlers and Shen 2017; Chinese State Council 2015; Tilt 2019). PM in mainland China has been skyrocketing, to the extent that everyday life cannot be sustained in many parts of the country—especially in Beijing, the northeast provinces, and the coastal provinces. This air pollution, along with other forms of severe environmental pollution, has even destabilized the Chinese authoritarian regime; since the early 2000s, the number of incidents of pollution-related social unrest has soared (Ahlers and Shen 2017; Nielsen and Ho 2013). Since Xi Jinping came to power in 2012, and particularly in the last few years, the Chinese central government has made great efforts to reduce PM across China, especially in the Beijing area (Chinese State Council

2015; Zhen 2018).[21] Even though such a strong drive is often labeled "environmental authoritarianism" (Beeson 2015) and involves penalizing polluting industries as well as disadvantaged citizens in order to reduce PM, particularly in winter (Ahlers and Shen 2017; Beeson 2015), it has vastly improved the region's air quality and was praised by domestic and international media (Zhen 2018). In fact, this success was lauded by many South Korean media outlets, which criticized their own government's inability to reduce domestically produced PM (Yang 2018). To be sure, despite this domestic success, PM within and outside of China remains a severe hazard.

Reducing asymmetries may facilitate the atmospheric politics of transboundary cooperation. It may be challenging to lessen state capacity asymmetry within the short term, but economic asymmetries can be reduced by creating win-win economic benefits for transboundary air pollution. Sharing best practices and technologies to mitigate PM in both countries can produce economic benefits. To this end, identifying cutting-edge, effective, and sharable air pollution reduction technologies should be prioritized. In addition, strengthening the international regime can lower asymmetries in transboundary air pollution. Countries in Northeast Asia have developed multilateral environmental cooperation networks, including the Acid Deposition Monitoring Network in East Asia (EANET) and Long-range Transboundary Air Pollutants (LTP) (Yarime and Li 2018). The current development (October 2018) of the Northeast Asia Clean Air Partnership (NEACAP) provides a multilateral arena as "a voluntary, science-based, and policy-oriented programme under NEASPEC" (UNESCAP 2018). In addition to bilateral efforts for cooperation, a multilateral regime can put forward potential legal agreements for transboundary air pollution by proposing technical and policy measures for Northeast Asian countries (UNESCAP 2018).[22] Bilateral cooperation between Korea and China would be facilitated through the existing multilateral efforts. Despite the challenges in multilateral coordination, LTP, EANET, and NEASPEC could reduce the competition over the concern on "who takes initiatives" as well as offer forum for collaboration.

In the process of regime building, lessons from EU CLRTAP, ASEAN agreement on transboundary haze pollution, and the Paris agreement on climate change should be acknowledged and reflected. Forming and operating multilateral cooperation in Europe for 40 years provide promising reference to build up effective atmospheric political cooperation (Fraenkel 1989; Tuinstra et al. 2006). Recent formation of ASEAN agreement could also offer how multiple participants make an agreement to curb regional transboundary air pollution. A close examination of the effectiveness of ASEAN agreement will guide East Asian countries' regime design and implementation scheme, avoiding a symbolic institution formation (Nurhidayah et al. 2015).

Particularly, a multilateral nationally determined contribution (NDC) agreement, utilizing the Paris Agreement format for climate change, would be a scheme to reduce transboundary air pollution in the region. In such a scheme, each country nationally commits a reduction target, makes plans, implements pollution reduction policies, and ratifies the agreement, conducting measure, report, and verify (MRV) pollutant reduction and capacity building support through international treaty organizations. In the situation of asymmetries, multilateral cooperation, instead of bilateral polluter's pay principle such as USA-Canada air pollution settlement case, would be a long-term, plausible, and comprehensive strategic solution. Furthermore, strengthening epistemic communities that share a core belief in causes and effects as well as policy options might help reduce asymmetries in knowledge and thus facilitate collaborative research.

5.4   Conclusion: implications and further studies

This study develops a conceptual framework for one of the key atmospheric political issues, transboundary air pollution, with an in-depth analysis of factors that hinder the development of cooperation to resolve issues related to transboundary air pollution between China and South Korea since the early 2010s. We discussed three asymmetric factors—state capacity, economic interest, and domestic pressure—and the element of international regime pressure as key variables hindering bilateral cooperation to reduce transboundary PM moving from China to South Korea. Empirical studies on transboundary air pollution cooperation thus far have primarily examined European and Southeast Asian cases; Northeast Asia has been understudied, with few exceptions (Kim 2007, 2014; Shim 2017). Our study fills this gap in the literature.

Considering these factors' negative effects on cooperation between the source country and the receptor country, it is clear that bilateral cooperation to resolve the transboundary PM issue between China and South Korea faces severe difficulties in the short term (Shim 2017).[23] Cooperation will likely be stuck at the second step of the suggested conflict-resolution model of collaborative research, even though the South Korean government continues to make efforts to negotiate with China bilaterally and multilaterally. Moreover, it might be problematic for South Korea to demand that China reduce the transboundary PM, or even to reduce its own PM production, because this pollution generation is also attributed to its economic interdependence with Chinese polluting industries. Moreover, South Korea produced a massive amount of PM during its industrialization period and, to a lesser extent, still does.[24] Unfortunately for South Korea, it happens to be on the downstream side of wind currents, while China is on the upstream.[25]

In this context, as discussed above, South Korea has no choice but to hope that domestic pressure in China to accelerate the reduction process of Chinese intranational transboundary PM is maintained and enforced. Reducing PM within China's own territory would automatically reduce the volume of transferred PM to South Korea, leaving China more open to cooperating with South Korea (and Japan) in the atmospheric political arena. Ironically, China's domestic pressure to protect its own citizens, who generally deny the harmful effects of transboundary PM on South Korea, will most likely serve as an impetus for China to address the PM problem that also affects South Korea. In addition, current initiatives for climate change and carbon neutrality would prevail, the NDC type of international air pollution cooperation could be realized. If this is the case, China, with its improving environmental credentials, is more likely to communicate and cooperate with South Korea on this issue in the coming years.

For students of global environmental politics, comparative politics, and China and Korea studies, addressing severe transboundary PM is an urgent task for both academic and policy purposes. Topics that will be addressed in further research are as follows. First, comparative studies between ASEAN's transboundary haze pollution agreement and cooperation efforts and those of Northeast Asian countries—China, South Korea, North Korea, Mongolia, and Japan—will be pursued to generalize the conceptual frameworks that were proposed in this paper. Second, we will focus on the sub-national and intra-national level of analysis by examining important domestic actors such as local governments, industries, media, NGOs, and ordinary citizens in atmospheric politics.

Notes

1 Please see air quality data at www.airkorea.or.kr/web/realSearch?pMENU_NO=97
2 As explained below, our analysis focuses on the bilateral case rather than the multilateral case, which needs more theoretical sophistication.
3 Note that financial support/subsidy transfer from the receiving country to the transferring country is another aspect of this economic-interest dimension. In fact, the case of China-South Korea experienced this economic interaction at a very low scale, which does not qualify as a game-changer due to the difference in size between the two countries and the sheer volume of air pollution in mainland China. If the receiving end is much larger economically and more capable of subsidizing the transferring end, this might be a critical factor in the game. This policy suggestion is applicable to the case of North Korea (source) and South Korea (receptor).
4 The degree of domestic audience pressure depends on multiple factors (variables) such as the nature of the political regime (democratic or authoritarian), the level of economic development (poor, developing, or developed), and scientific information on the various features of the air pollution (transboundary, domestic-generated, atmospheric dynamics, impact on quality of life, etc.), which are beyond the scope of this paper.
5 Note that in addition to the interviews conducted during the fieldwork, some interviews were conducted via email and phone to obtain more diverse interviewees in China and South Korea.
6 July 3, 2014, Memorandum of Understanding between the Ministry of Environment of the Republic of Korea and the Ministry of Environmental Protection of the People's Republic of China on Environmental Cooperation; 2016 MOU for Collaborative Research Team on Air Quality of Korea and China.
7 Interviews with South Korean experts, January 12, August 23, 2019; officials January 13, 2019. This might be political reasons to avoid the responsibility.
8 Despite a series of research efforts inside and outside of China, we could not find any data that reported the scientific evaluation of transboundary air pollution between China and South Korea. See Kim and Kim (2018).
9 Hangyoreh Newspaper, June 19. 2018. China, Japan, and Korea's collaborative publication of PM from China broken down. (www.hani.co.kr/arti/society/environment/849781.html

10 Interview with Chinese expert, January 11, 2019; Chinese official, January 20, 2019. Cho (2019).

11 Data retrieved from the World Bank Dataset (https://data.worldbank.org/indicator/NY.GDP.MKTP.CD) and SIPRI dataset (www.sipri.org/databases/milex).

12 Data retrieved from the Korea Customs dataset (https://unipass.customs.go.kr:38030/ets/).

13 Interviews with South Korean officials and experts, January 12, 2019; Chinese experts and journalist, January 8, 11, and 25, 2019.

14 Interviews with Chinese experts and officials, January 8, 11, 20, and 25, 2019

15 Interviews with Chinese experts and officials, January 8, 11, 20, and 25, 2019.

16 Interview with South Korean experts, January 13, 2019.

17 By contrast, North Korea is a closed authoritarian state with an underdeveloped economy and little scientific knowledge. North Korea is not demanding that China settle this transboundary PM issue.

18 Interviews with Chinese experts, January 8 and 25, 2019.

19 Interviews with Chinese experts, November 16, 2018; January 9, 11, 20, and 25.

20 Joint research projects (of China, Japan, and South Korea) for Long-range transboundary air pollutants in Northeast Asia recently announced the summary report of 4th stage (regarding PM) LTP project (2019. Nov. 20). LTP (2019). While it was the first collaborative research outcome announcement, detailed contents on the influence of sources (Chinese cities) on receptors (South Korean cities) between China and South Korea estimates are not identical due to data gap. In addition, transports of PM during high level seasons (from December to March) were not included in the report (Kim 2019).

21 Interviews with Chinese experts, November 16, 2018; January 9, 11, and 17, 2019.

22 Interviews with a South Korean expert, January 21, 2019.

23 Interviews with Chinese experts, January 11, 2019; Chinese officials, January 20, 2019; South Korean experts, January 12, May 21, 2019.

24 It is analogous to the certified emissions reduction issue in climate-change debate, because South Korea produced a massive amount of PM in the last several decades without being recognized as a source-transferor to its neighboring countries, but now blames China for generating and transferring PM to it.

25 Hypothetically, if China were downstream and South Korea were upstream, there would be a much lower level of conflict revolving around the transboundary PM. Instead, China might be in South Korea's position and be demanding a resolution to the issue. If so, South Korea would be under severe pressure to cooperate with its more powerful neighbor, and the asymmetric logic would run the opposite way, accelerating the cooperation process. From a different perspective, the primary receptors victimized by the massive PM transfer from China would be the countries of Central Asia and Southeast Asia.

References

Ahlers, A. L., and Shen, Y. (2017). Breathe Easy? Local Nuances of Authoritarian Environmentalism in China's Battle against Air Pollution. *The China Quarterly*, 234, 299–319.

Beeson, M. (2015). Authoritarian Environmentalism and China. In T. Gabrielson, C. Hall, J. M. Meyer, and D. Schlosberg (Eds.), *The Oxford Handbook of Environmental Political Theory* (pp. 520–532). Oxford: Oxford University Press.

Chinese Ministry of Foreign Affairs. (2017). 2017 nian 3 yue 21 ri waijiaobu fayanren huanchunying zhuchi lixing jizhehui. *Ministry of Foreign Affairs Spokesperson*

Hua Chunying Conducts a Press Conference, March 21, 2017. www.fmprc.gov.cn/
web/fyrbt_673021/jzhsl_673025/t1447426.shtml, accessed 21 March 2018.

Chinese State Council. (2015). *Shengtai Wenming Tizhi Gaige Zongti Fangan*.
General Scheme for the Reform of the Ecological Civilization System, Beijing.

Cho, Arum. (2019). Minister of Environment, Myoungrae Cho Say, "Chinese
Ministry of Ecology and Environment Acknowledges the PM from China". *Han-
kook Ilbo*, March 7, 2019. www.hankookilbo.com/News/Read/2019030715773
54424?did=DA&dtype=&dtypecode=&prnewsid=, accessed March 21, 2019.

Daoudy, M. (2009). Asymmetric Power: Negotiating Water in the Euphrates and
Tigris. *International Negotiation*, 14(2), 361–391.

Dinar, S. (2009). Power Asymmetry and Negotiations in International River Basins.
International Negotiation, 14(2), 329–360.

Dong, L. (2017). Wu mai zeren, huanjing waijiao yu zhongrihan hezuo. Responsibility for PM, Environmental Diplomacy and China-South Korea Cooperation.
Dangdaihanguo. *Contemporary Korea*, 2, 1–14.

Dong, X. (2018). Hanguo de wumai laizi Zhongguo? Tamen ziji de! [South Korea's
PM Comes from China? It Comes from Itself!] *Beijingqingnianbao. Beijing Youth
Daily*, December 28. www.sohu.com/a/285136718_255783, accessed December 6,
2019.

Fraenkel, A. A. (1989). The Convention on Long-Range Transboundary Air Pollution: Meeting the Challenge of International Cooperation. *Harvard International
Law Journal*, 30(2), 447–476.

Gourevitch, P. (2002). Domestic Politics and International Relations. In W. Carlsnaes,
T. Risse, and B. A. Simmons (Eds.), *Handbook of International Relations* (pp. 309–328).
SAGE Publications Ltd, https://www.doi.org/10.4135/9781848608290.n16

Gu, Y., and Yim, S. H. L. (2016). The Air Quality and Health Impacts of Domestic
Trans-boundary Pollution in Various Regions of China. *Environment International*,
97, 117–124.

Haas, P. M. (1990). *Saving the Mediterranean: The Politics of International Environmental Cooperation*. New York: Columbia University Press.

Han, H. (2017). China, an Upstream Hegemon: A Destabilizer for the Governance
of the Mekong River? *Pacific Focus*, 32(1), 30–55.

Harris, P., and Lee, T. (2017). Compliance with Climate Change Agreements:
The Constraints of Consumption. *International Environmental Agreement*,
17(6), 779–794. Huanqiushibao. (2017). *Hanguoren yin shouer wumai yao Beijing peichang jingshen sunshi, duicini zenmekan?* [How Do You Think about
South Korean Demanded Beijing Reparation in Cash for Their Mental Damage
by PMs?]. http://surveyx.huanqiu.com/survey/survey_hq/68, accessed April 6,
2018.

Hussein, H., and Grandi, M. (2017). Dynamic Political Contexts and Power Asymmetries: The Cases of the Blue Nile and the Yarmouk Rivers. *International Environmental Agreements: Politics, Law and Economics*, 17(6), 795–814.

Keohane, R. O., and Joseph, N. (2012). *Power and Interdependence*, 4th ed. Boston:
Longman.

Kim, H. (2019). 32% of PM Comes from China. China First Admitted. *Chosun Ilbo*.
https://news.chosun.com/site/data/html_dir/2019/11/21/2019112100269.
html, accessed December 6, 2019.

Kim, I. (2007). Environmental Cooperation of Northeast Asia: Transboundary Air
Pollution. *International Relations of the Asia-Pacific*, 7(3), 439–462.

Kim, I. (2014). Messages from a Middle Power: Participation by the Republic of Korea in Regional Environmental Cooperation on Transboundary Air Pollution Issues. *International Environmental Agreements: Politics, Law and Economics*, 14(2), 147–162.

Kim, S. (2018). Sijinping ege masuku batdawara. jaenangup misemeonjie gyukyang [Make Xi Jinping Pay for Dust Masks. Soaring Anger Caused by Disastrous Particulate Matter [PM]]. *Newsis*. www.newsis.com/view/?id=NISX20180326_0000262407, accessed December 5, 2019.

Kim, S., and Kim, D. (2018). Analysis of the Recognition Differences and Cooperation between Korea and China on the Transboundary Environmental Pollution Issues: Focusing on Peaceful Conflict Resolution (in Korean). *The Journal of Peace Studies*, 19(1), 253–277.

Lee, S., Kim, J., Choi, M., Eck, T. F., Hong, J., Lim, H., et al. (2019). Analysis of Long-range Transboundary Transport (LRTT) Effect on Korean Aerosol Pollution during the KORUS-AQ Campaign. *Atmospheric Environment*, 204, 53–67.

Liu, Q., and Wang, Q. (2017). Zhongrihan kuajie daqiwuranzhong de zhongguo zeren shibie yanjiu [Study on China's Responsibility in the China-Japan-South Korea Transboundary Atmospheric Pollution]. *Dongbeiyaruntan [Northeast Asia Forum]*, 6, 7–91.

LTP. (2019). *Summary Report of the 4th Stage (2013–2017) LPT Project*. NIER: Joint Research Project for Long-range Transboundary Air Pollutants in Northeast Asia.

Luterbacher, U., and Sprinz, D. F. (Eds.). (2001). *International Relations and Global Climate Change*. Cambridge, MA: MIT Press.

Min, B. (2001). Regional Cooperation for Control of Transboundary Air Pollution in East Asia. *Journal of Asian Economics*, 12(1), 137–153.

Mitchell, Ronald B. (2010). *International Politics and the Environment*. California: Sage Publications.

MoE. (1993). *Agreement on Environmental Cooperation Between the Government of the Republic of Korea and the Government of the People's Republic of China, ROK-China*, October 28, 1993. National Archives of Korea.

MoE. (2003). *Memorandum of Understanding between the Ministry of Environment of the Republic of Korea and the State Environmental Protection Administration of the People's Republic of China on Environmental Cooperation, ROK-China*, July 8, 2003. Ministry of Environment.

MoE. (2005). *Arrangement between the Ministry of the Environment of the Republic of Korea and the State Environmental Protection Administration of the People's Republic of China on Ground Monitoring and Information Exchange for Dust and Sand Storms, ROK-China*, June 7, 2005. Ministry of Environment.

Nielsen, C. P., and Mun, S. H. (2013). *Clearer Skies Over China: Reconciling Air Quality, Climate, and Economic Goals*. Cambridge, MA: MIT Press.

NIER (National Institute of Environmental Research) (2017). *KORUS-AQ Rapid Science Synthesis Report*. NIER.

Nurhidayah, L., Shawkat, A., and Lipman, Z. (2015). The Influence of International Law upon ASEAN Approaches in Addressing Transboundary Haze Pollution in Southeast Asia. *Contemporary Southeast Asia: A Journal of International and Strategic Affairs*, 37(2), 183–210.

Oh, H., Ho, C., Kim, J., Chen, D., Lee, S., Choi, Y., et al. (2015). Long-range Transport of Air Pollutants Originating in China: A Possible Major Cause of Multi-day

High-PM10 Episodes During Cold Season in Seoul, Korea. *Atmospheric Environment*, 109, 23–30.

Oh, J. H. (2019). Summakhi Han . . . Misemunji, Junge halmaleun hara [Suffocated South Korea . . . Say Its Piece to China] *Seoul Economic Daily* Page 1, 4, 5, January 15.

O'Neill, K. (2009). *The Environment and International Relations*. New York: Cambridge University Press.

Park, S. M. (2019). 2020 yesan: misaemunji haegyeole sajo ssenda . . . muknunmul gwanlido ganghwa [2020 Budget: Spending 4 Trillion Korean Won to Settle the PM Pollution . . . and Enforcing the Tap Water Management]. *Yeonhapnews*, August 29. www.yna.co.kr/view/AKR20190828157800004, accessed December 3, 2019.

Perrin, S., and Bernauer, T. (2010). International Regime Formation Revisited: Explaining Ratification Behaviour with Respect to Long-range Transboundary Air Pollution Agreements in Europe. *European Union Politics*, 11(3), 405–426.

President Office. (2018). *Citizen Petition on Protest Against Particulate Matter from China*. http://www1.president.go.kr/petitions/174292

Putnam, R. (1988). Diplomacy and Domestic Politics: The Logic of Two-level Game. *International Organization*, 42(3), 427–460.

Roelofs, Jeffrey. (1993). United States-Canada Air Quality Agreement: A Framework for Addressing Transboundary Air Pollution Problems. *Cornell International Law Journal*, 26(2), 421–454.

Selin, H., and VanDeveer, S. (2003). Mapping Institutional Linkages in European Air Pollution Politics. *Global Environmental Politics*, 3(3), 14–46.

Shim, C. (2017). Policy Measures for Mitigating Fine Particle Pollution in Korea and Suggestions for Expediting International Dialogue in East Asia. *JICA-RI Working Paper*, pp. 1–31. https://www.jica.go.jp/jica-ri/publication/workingpaper/wp_150.html

Shin, G., Izatt, H., and Moon, R. J. (2016). Asymmetry of Power and Attention in Alliance Politics: The US—Republic of Korea Case. *Australian Journal of International Affairs*, 70(3), 1–21.

Stephen, P. (2009). Asymmetry and Selectivity: What Happens in International Law When the World Changes. *Chicago Journal of International Law*, 10(1), 91–123.

Tilt, B. (2019). Review: China's Air Pollution Crisis: Science and Policy Perspectives. *Environmental Science and Policy*, 92, 275–280.

Tuinstra, W., Hordijk, L., and Kroeze, C. (2006). Moving Boundaries in Transboundary Air Pollution Co-production of Science and Policy under the Convention on Long Range Transboundary Air Pollution. *Global Environmental Change*, 16, 349–363.

UNESCAP. (2018). *Review of Programme Planning and Implementation Transboundary Air Pollution in North-East Asia: Twenty-second Senior Officials Meeting (SOM) of NEASPEC*. Bangkok: UNESCAP.

Warner, J., and Zawahri, N. (2012). Hegemony and Asymmetry: Multiple-chessboard Games on Transboundary Rivers. *International Environmental Agreements: Politics, Law and Economics*, 12, 215–229.

Womack, B. (2015). *Asymmetry and International Relationships*. Cambridge: Cambridge University Press.

Yang, G. (2018). Misemeonji wayi jeonjaeng junggukeu sunggonghanunde [China Succeeds in War against PMs]. *Sisain*. www.sisain.co.kr/?mod=news&act=articleView&idxno=31197, accessed Februrary 7, 2018.

Yarime, M., and Aitong, Li. (2018). Facilitating International Cooperation on Air Pollution in East Asia: Fragmentation of the Epistemic Communities. *Global Policy*, 9(S3), 35–41.

Yoon, E. (2007). Cooperation for Transboundary Pollution in Northeast Asia: Non-binding Agreements and Regional Countries' Policy Interests. *Pacific Focus*, 22(2), 77–112.

Zhen, L. (2018). Beijing Meets National Air Pollution Standard for First Time. *South China Morning Post*. www.scmp.com/news/china/society/article/2132406/beijing-meets-national-air-pollutant-standardfirst-time, accessed Februrary 7, 2018.

Appendix 1
List of interviewees

Interviews were conducted in a face-to-face manner with semi-structured questions concerning the causes or drivers of international cooperation over transboundary air pollution. Interviewees agreed to be anonymously interviewed. All interviews lasted between thirty minutes and one hour.

Code	Background	Date
1	Professor at a University in Northeast China (International Relations)	8 & 11 & 20 January 2019
2	Researcher at a University in South Korea (China Studies, China-South Korea Relations)	16 November 2018; 25 January 2019
3	Journalist from a Newspaper in Beijing	25 January 2019
4	Local government official from Tianjin (External Affairs)	25 January 2019
5	Chinese government official from the Ministry of Commerce	24 January 2019
6	Scholar from a University in Beijing	26 January 2019
7	Chinese scholar from a University in Seoul	9 & 17 January 2019
8	Local government official from Jilin (Environmental Affairs)	20 January 2019
9	Professor of a University in Jilin Province (China Studies, China-South Korea Relations)	20 January 2019
10	South Korean official (Foreign affairs)	9 May 2019
11	Researcher at Hanyang University, South Korea (Environmental Politics)	23 August 2019
12	Professor from Korea University (Energy and Environmental Politics)	12 August 2019
13	Professor from Kwangwoon University, South Korea (Environmental Politics)	4 July 2019
14	Environmental activist from South Korea	4 July 2019
15	Professor of Atmospheric science at Yonsei University	21 May 2019
16	Researcher from the Korea Environmental Institute	21 May 2019

6 Atmospheric environment management regime building in East Asia

Limitation of imitating the convention on long-range transboundary air pollution and current new development

Jusen Asuka

6.1 History of transboundary air pollution in the East Asian region

The history of North-East Asia region, including Japan, China, and Korea, shows several transboundary air pollutants drawing attentions intermittently.

6.1.1 Aerial flow of yellow dust reported since the age of Man-Yo poetries (mid-seventh century)

The first of East Asian transboundary air pollutants was yellow dust (mineral dust). Annual amounts of dust generated from the source areas are estimated to be 400–500 million tons with 200–300 million tons being re-deposited in the source areas and about 100 million tons being transported over long distances (Xuan et al. 2004).

Since the phenomenon mostly happens in spring time from March to May, it used to be called "Spring Haze (Haru Gasumi)" in Japan. The phenomenon itself has been known from 1000 or more years ago. In fact, one of the poetry books called Man-Yo-Shu (published around 760 AD) had a poetry themed "Haru Gasumi." The origin of yellow dust is said to be Taklamakan Desert, Gobi Desert, and Huang Tu Plateau. Basically, natural phenomenon, but in view of desertification being anthropogenic. It can be called natural phenomenon with anthropogenic elements.

6.1.2 Becoming diplomatic issue

In the latter half of the twentieth century, transboundary air pollutions became national diplomatic issues. One big factor behind such development was greater focus being given to global and/or regional environmental issues, as a new political issue emerged through the progress of globalization, filling the void

DOI: 10.4324/9781003211747-8

vacated by the end of the Cold War (Yonemoto 1994). Moreover, the significant development of scientific researches using air dispersion models and other modeling studies has had considerable effects, as this paper has highlighted.

The first time this transboundary air pollutions became the "diplomatic issue" was the transboundary acid rains in European countries during the 1970s (Wettestad 1996; Levy 1993). Officially, this acid rain issue is "the issue of acid rain over ecosystem," in which air pollutants, such as sulfur oxides mainly from the combustion of coals and nitrogen oxides in the exhaust gas of automobiles, causes the destruction of forests, less yields and less income from agricultural products, the damages to buildings, etc.

In Europe, the scientific studies reported clearly since the 1970s that the emissions of air pollutants from the sources in Western and Eastern Europe, including factories and power stations, transferred to wide area, impacting considerable effects upon the ecosystems of North Europe (Amann et al. 1992; EEA 1997; Bjorkbom 1999). These studies indicated that about 80–90% of deposited quantities of air pollutants in Northern European countries were from outside sources. As the emission reduction in domestic sources would not have sufficient effects to reduce acid rain, Northern European countries requested Western and Eastern European countries to reduce emissions, bringing it up a big diplomatic issue. Their actions resulted in increased interests in transboundary acid rain in North American and Asian regions.

At that time, the East Asian region, especially the southern part of China, had experienced serious acid rain damages. The overwhelming cause of acid rain was the coal consumption in China, especially the use of coals with higher sulfur contents. Japan started to argue about the relationship between the frequently seen deadwood of pine and cedar trees and the transboundary acid rain from China.[1] Korea raised similar concerns over the effects of air pollutants from China.

With the rising concerns, the Ministry of Environment Japan (MOEJ) conducted a large-scale survey for acid rain damages for several years. In the fourth acid rain measures survey announced in 2002, MOEJ disclosed the result of the survey and indicated that there was no soil acidification to suggest a clear relationship with the deposit of acidic materials and no apparent damages of acid rain found in Japan. Based on this result, the Japanese public lost interest in transboundary acid rain issues.

On the other hand, the acid rain issue led to the fixation of Japanese public's awareness about "many air pollutants to fly from China." In fact, there were many kinds of air pollutants transported from a wide area, in addition to sulfur oxides and nitrogen oxides. Thus, the focus of transboundary air pollutants shifted from acid rains to yellow dust, and in the late 2000s to ozone (O^3), which caused photochemical smog. After 2013, when air pollution of $PM_{2.5}$ became a topic of the day in China, $PM_{2.5}$ as a transboundary air pollutant drew attention in East Asian nations.

There is a significant difference in terms of the implications between the problems of "acid rain and yellow dust" and the problems of "ozone and $PM_{2.5}$."

The former mostly concerns the adverse effects on the ecosystem as a whole. The latter, on the other hand, concerns the adverse effects on human health and health damages. In this sense, the latter problems can be described as much more serious ones.

Still, the transboundary pollution of $PM_{2.5}$ has not reached the stage of a big diplomatic issue in Japan, yet. Japanese government officials have not criticized their counterparts in public places. However, there have been some occasions where diplomats have criticized each other publicly between China and Korea with more Chinese influences. For example, during the regular press conference held by the Chinese Ecological Environmental Department in Beijing on December 28, 2018, the Managing Deputy Chief (Leading Deputy Director) of the Central Environmental Protection and Inspection Office responded to the question asked by the media, saying that "although Korea claims trans-boundary pollutions from China, major sources of pollutants are from domestic sources" (Koyanagi 2019). Korea immediately responded to this with their government official saying something like "China tends to interpret everything to their own advantages." According to the Korean Daily Report, Mr. Cho Myung-le, Director-General of Korean Environmental Department did exactly state on January 3, 2019, that "China tends to interpret everything to their own advantages," while Mr. Park Won-soon, Mayor of Seoul City, said on January 7, 2019, that "plural number of studies show that more than 50% or 60% of particulates are the effects of China" (Koyanagi 2019).

6.2 Scientific infrastructure and political infrastructure

The long-range transportation (dispersion) of air pollutants beyond national borders is a proven scientific fact. In order to change such a situation through international negotiation, it is essential to quantify the volume of such long-range transportation and to identify the responsibilities of relevant countries. For this purpose, what is required is to prove concrete "evidence in science." At the same time, it is necessary to develop a political forum to discuss countermeasures and a framework for technological and financial cooperation. In other words, there should be a regime for the control of atmospheric environment to institute both scientific infrastructure and political infrastructure (Yonemoto 1994).

6.2.1 LRTAP

In 1979, the United Nations Economic Commission for Europe (UNECE) adopted the Convention on Long-Range Transboundary Air Pollution (LRTAP), which entered into force in 1983. EU and 51 other nations have become the parties to this convention. To agree on LRTAP, European Monitoring and Evaluation Program (EMEP), implemented since the 1970s, had an important role.

LRTAP was the first international regime for the control of atmospheric environment to address transboundary air pollution problems, and its regulations subjected various transboundary pollutants in Europe and North America. After its adoption in 1979, the Convention had eight Protocols to compliment and reinforce the regulatory subjects.

The notable features of LRTAP are that a third-party research institute is to quantify how transboundary pollutants influence different nations, and the burden of responsibilities of each nation is to be allocated among nations on a scientific basis founded on the result of such quantification.[2] While promoting technological and financial cooperation among parties, the air pollutants subjected to the regulations of LRTAP have been expanded step by step, and the parties set up the obligations to realize actual emission reductions.

6.2.2 Infrastructure building

As stated above, in order to clarify the cause-result relationships and in the occurrences of damages from the wide-ranged transportation of air pollutants, to identify their international effects, and to enforce actual measures through international cooperation, it is necessary to establish scientific infrastructure to: 1) implement the quantifiable monitoring of damages from acid rain; 2) build integrated institution to enable each party to collect monitoring data, to compare, mutually, and to make assessment, etc.; 3) get agreement on the result of long-range dispersion model study to indicate the effectual relationships between parties. Moreover, it is essential to establish political infrastructures for common policies, technology transfer, and fund transfer.

As stated, LRTAP already presented a grand design in the 1970s to establish not only "scientific infrastructure" but also "political infrastructure."

6.3 Regime in East Asia to control transboundary air pollution

For Japan and Korea located under the wind and gotten damages of air pollution, it was inevitable and natural to consider it ideal to establish an LRTAP-like regime to control the atmospheric environment in East Asia. With EMEP and LATAP in reference, Asian countries attempted to establish "scientific infrastructure" first, and then to build "political infrastructure" simultaneously while examining the situation.

6.3.1 Acid rain monitoring network in East Asia

In a way following the path of Europe, Asian countries finally established a basic agreement on the establishment of integrated institution in 1997. That agreement created by the initiative of the Japanese government was called the East Asian Network for the monitoring of acid rain (EANET). Its parties include

China, Indonesia, Japan, Malaysia, Mongolia, Philippines, Republic of Korea, Russia, Thailand, and Viet Num.

Its Secretariat is located in Niigata Prefecture of Japan, offering trainings of researchers from EANET parties as capability-building activities. Japan contributes almost all of the funds to operate EANET, which in a way encourages the participation of regional countries.

The Network provides a framework of multilateral cooperation in environmental fields for the first time in the Asian region, under which academic researchers are expected to provide policy options and data availability to policymakers.

However, the Network has a limited range of activities in monitoring and monitoring-related research and survey. Therefore, it has failed to send out a strong message that might lead to the actual reduction of air pollutants.

6.3.2   *Joint research project to study long-range and transboundary transport of air pollutants in North-eastern Asian region*

A joint research project to study long-range transboundary transport of air pollutants (LTP) started under the initiative of Korea. Only three countries of Japan, China, and Korea are participating in this project. In September 1995, the first workshop to study long-range transboundary transport of air pollutants in the Northeastern Asian region was held in Seoul, which decided two items of: 1) implementation of expert meetings; and 2) implementation of joint studies under the expert meetings. As a result of this workshop, LTP was established and started to undertake various projects since 1999.

The characteristics of LTP are, unlike EANET, to stipulate the implementation of joint researches for the assessment of air pollution effects and long-range dispersion modeling, as some of its main activities. On the other hand, LTP does not have an organizational structure like EANET, which has Secretariat with many office staff.

6.4   Contribution of emission sources to air pollution in East Asia

The model of long-range dispersion of air pollutants clearly indicated the numerical value (or contribution ratio of sources) of source-receptor relationship, that is, source region and receptor region relationship to indicate the effects relationships between countries. The numerical value is extremely important in identifying the responsibilities of regional nations over transboundary pollution, which shows how the emissions of air pollutants in one nation affect the concentration of such pollutants in another country.

As mentioned above, LRTAP was formulated under strong initiatives of Nordic countries, where scientific research identified that the contribution ratio of domestic air pollutant emissions in acidic materials deposits was less than 10 percent.

Stated below are the major results of modeling study calculations for the source-receptor relationships of air pollutants, which used to present problems in East Asia.

6.4.1 *Source-receptor relationships of transboundary acid rain*

The past studies clearly indicated that there were three emission sources of sulfur oxides deposited in the atmosphere and ground surface of Japan: anthropogenic emissions from factories, power stations, automobiles, etc., in Japan; volcanic emissions from Japanese volcanoes (mostly from Mt. Sakurajima in Kyushu Island); and those emissions flew from Chinese continent.

As shown in Table 6.1, during the latter half of the 1990s when transboundary acid rain became a big problem, the researchers in Japan, Europe, and the USA were sharing the common perception that "sulfur emissions from China would be transported to Japan, and the Chinese continental sulfur would contribute to 10 to 50% of sulfur deposits in Japan." Chinese researchers, on the other hand, claimed that "Chinese continental sulfur would contribute to only several percent of sulfur deposits in Japan," presenting one digit less number. Thus, in the case of acid rain, the opinions of Japanese researchers and Chinese researchers differed significantly.

Table 6.1 Estimates on the contribution of each sulfur emission source to sulfur depositions in Japan (%)

Research Group Name (Projection providers)	Target Year	Type of Models used	Emission Source				
			Japan	Volcanic	China	Korean Peninsula	Others
Electric Power Central Research Committee (Ichikawa, Fujita, Hayami)	1988.10–1989.09	Hybrid btw Trajectory and Oiler	40	18	25	16	1
Osaka City University (Ikeda, Touno)	1990	Oiler type	76	–	13	11	0
Yamanashi University (Kataya)	1988	Oiler type	47	11	32	10	0
IIASA-RAINS ASIA (Carmichael, Arndt)	1989	Trajectory type	38	45	10	7	0
Chinese Science Committee (Huang, et al)	1989	Oiler type	Japan and volcanic together being 94		3	2	1

Source: Ichikawa (1998)

6.4.2   *Source-receptor relationships of ozone*

After acid rain, ozone drew attention as a material that causes health damages. In regards to the flow of ozone from China, Japanese research institutes, such as Japan Agency for Marine Earth Science and Technology, and National Institute for Environmental Studies, have done quantitative analysis. For example, Nagashima et al. (2010) indicated that Chinese origin ozone in the ozone concentration over Japan was about 12% in spring time, and about 19% in summer.

6.4.3   *Source-receptor relationships of* $PM_{2.5}$

In regards to $PM_{2.5}$ flying from China, Japanese researchers have conducted quantitative analysis as well. For example, Ikeda et al. (2014) found that in the area west of Kanto Plain, the ratio of contribution to $PM_{2.5}$ concentration was 20–50% from domestic sources, 0–10% from Korean Peninsula, and 40–50% from China.

For $PM_{2.5}$, the joint research under LTP, for which researchers from Japan, China, and Korea participated, published their conclusions recently. Under LTP, the researchers of three countries studied the atmospheric concentrations of $PM_{2.5}$ in major cities (China: Beijing, Tianjin, Shanghai, Qingdao, Shenyang and Dalian/Korea: Seoul, Daejeon and Busan/Japan: Fukuoka, Osaka, and Tokyo), and calculated the estimated ratios of contribution to $PM_{2.5}$ concentration for emission sources in Japan, China, and Korea, through simulations using atmospheric dispersion models of each country. The LTP used the differences in contribution ratios for their discussion and negotiation.

Table 6.2 shows the latest result of the above studies for the fourth term (2013 to 2017) announced in January 2020. When using the data of 2017, the contribution ratios of domestic sources toward $PM_{2.5}$ concentration in major cities were 91.0% for China, 51.2% for Korea, and 55.4% for Japan, on average

Table 6.2 Contribution rate of each country on $PM_{2.5}$ concentration by LTP study

Country	Contribution rate
China	From China: 91.0% From Korea: 1.9% From Japan: 0.8%
Korea	From China: 32.1% From Korea: 51.2% From Japan: 1.5%
Japan	From China: 24.6% From Korea: 8.2% From Japan: 55.4%

Source: MOEJ (2019)

Note: Since there exist other emitting sources, the total is not 100%.

of three models. Furthermore, $PM_{2.5}$ concentrations mutually affected those of three countries and the contribution ratios of Chinese sources were 32.1% in major Korean cities, and 24.6% in major Japanese cities, of Korean sources were 8.2% in major Japanese cities and 1.9% in major Chinese cities, and of Japanese sources were 0.8% in major Chinese cities and 1.5% in major Korean cities (MOEJ 2019).

The result of this LTP's latest studies showed a kind of consensus among government officials and researchers of three countries, so that this may put an end to the unconstructive discussion in the past.

6.4.4   Notes on the result of simulation studies

To interpret the result of simulation studies mentioned above, it is necessary to note the following:

Firstly, there is no "right" interpretation of the largeness of numerical values, as far as the model calculation of national contribution ratios is concerned. This is because the largeness of numbers is mostly a subjective matter. Some people may consider a certain number being large, and others may consider it small.

Secondly, the contribution ratios of air pollutants vary depending on the season. They also vary depending on the areas. In large cities, the contribution ratios of national emissions become higher especially due to the exhaust gas emissions from automobiles. In the case of the Tokyo metropolitan region, the contribution ratio of domestic sources is higher than that of overseas sources.

6.5   Difficulties of establishing a regime in East Asia

Building international agreement in a transboundary environmental issue is not easy, as relevant countries have conflicts of interest. In this sense, LRTAP is a rare success, as it is built with European countries at the center. When considering the establishment of a framework to address the transboundary issue in Asia, therefore, it is necessary to analyze objectively the differences between Asia and Europe as stated below.

6.5.1   Differences between East Asia and Europe

Firstly, Europe and Asia have different regional governances. In the case of EU, their "problem countries" are mainly Central and East European countries. When expanding the European Union to those countries, EU have had both canes and carrots for them, that is, canes as more strict environmental standards, and carrots as financial assistance for environmental conservation investments. Such measures can be construed as the issue linkage between environmental issues and political and economic issues.

In the case of Asian countries, there is almost no incentive to unify policies, when compared with EU. When creating an institution for international cooperation, especially inter-governmental cooperation, Asian nations will likely struggle to determine role sharing, and the order of priorities. The Asian region has vast diversities in terms of: political and economic systems, stages of economic development, cultural and social backgrounds, political intentions, perception of human rights including environmental rights, awareness of environmental issues needing urgent measures, etc. Thus, Asian countries have much greater difficulty in developing unified policies and measures.

In the case of Japan, its Official Development Aids (ODA) for China are mostly loans, which decreased noticeably since 2000 (Japan's ODA to China ended in 2019). So there no financial resources exist anymore which can be used as "carrots."

Secondly, the researches in the Asian region are not necessarily aiming for policy proposals. In Europe, researchers and research institutes continuously respond to political requests and provide data and policy options as a third party. In Asia, relatively speaking, there exist a certain gap between science and politics. Therefore, to make concrete policies, both science and politics need to make some compromises.

Moreover, in the case of LRTAP, it has designated International Institute for Applied Systems Analysis (IIASA) in Austria as a neutral research institute. In East Asia, on the other hand, it will be extremely difficult to make the designation of such an independent research institute.

Thirdly, there is the presence of China, which is a developing country in terms of per capita volume, but "a big country" in terms of various aspects including the quantity of air pollutant emissions. In Europe, there is no such nation emitting much more than others.

Fourthly, the contribution ratios of East Asian countries are not so strongly correlated as in the case of Europe. As stated above, in the case of transboundary acid rain, about 80–90% of air pollutant deposits in Nordic countries were from the sources in other countries, while in East Asia, the contribution ratios of foreign sources were relatively smaller.

Fifthly, there was not much manifestation of serious damages from transboundary acid rain, at least in Japan, when European countries started to raise such issues. Only from the perspectives of improving the environment of their own nation, these Asian countries can realize the importance of international cooperation and place higher political priorities in solving such environmental issues. These features have mutual effects upon the abovementioned four elements.

As a result, neither EANET nor LTP have not been developed into a Convention (Agreement) for the emission reduction of pollutants, like those seen in the Mediterranean region (for pollution of oceans) or in Europe (for transboundary acid rain).

6.5.2 *Political conflicts and political uses*

In the case of East Asian countries, political conflicts among them affect the building of any kind of cooperative regime. Diplomatic relationships among Japan, China, and Korea are not always amicable for the last several years, due to the conflicts in various issues. There have been some ups and downs, but such diplomatic relationships affected the proceedings of joint researches, etc. in some cases.

Especially, the rivalry relationship between Japan and Korea presented considerable effects. This was the real cause of disputes over which country was to take the leadership in building a regime. That was why Korea took its own initiative to build LTP, a multilateral cooperative framework like Japan-led EANET. So, LTP and EANET had a conflictive relationship at one time.

Moreover, it is evident that politicians of Japan and Korea are attempting to use such environmental issues to their advantage in diplomatic negotiation. Blaming other country's responsibility can provide the effects of diverting the interests of their public away from domestic problems and countermeasures.

On the other hand, these countries never canceled any meetings of environmental cooperation, even when their summit meetings or dialogue could not be held due to national conflicts over historical problems, etc. That means that, during governmental conflicts, environmental cooperation can become the only diplomatic pipeline that remained.

6.5.3   *Limitations in cooperation*

The problem is whether Japan, Korea, and China can cooperate with each other, and if they can, what kind of cooperation will be possible.

In the case of cooperation between Japan and China, for example, Japan has already transferred to China for a certain degree, its experiences of pollutions in the 1960s and 1970s as well as the knowhow on countermeasures through capacity-building and awareness-raising activities and projects, using technological cooperation under Official Development Aids (ODA). Japan's ODA to China ended in 2019, with general grants ending in 2006 and new Yen loans ending in 2007. Therefore, as stated previously in this article, it is difficult to procure Japanese public funds for financial and technological cooperation with China.

There are considerable misunderstandings about technological cooperation. Such as: People who do not understand the current situation say things like, "If the technology were transferred, the problem would be solved." Technology transfers are, in the end, commercially viable business transactions, and they will not happen without proper financial incentives for both sides. Moreover, nearly no special environmental technology exists these days that only Japan possesses. Companies around the world are marketing their technology fiercely to China, and the sophistication of Chinese firms is on the rise.

Recently, the Ministry of Environment Japan had about one billion JPY (9800 mil. US$) budget for Japan–China city-to-city cooperation for improving the atmospheric environment of China, under the five-year plan started in 2014 (Koyanagi 2019). Although Japan continues to implement such international cooperation, its scale is not large enough and hardly provides significant and immediate effects on the decrease of $PM_{2.5}$ concentration in China

6.5.4   *Self-help efforts of China*

There are already measures being taken in China that would be unimaginable in the western developed countries such as Japan. For example, a major social issue arose a few years ago when several local authorities in China forcibly cut electricity and gas supplies at certain times to meet energy conservation targets. And as for automobiles, Beijing has a lottery to obtain a license plate (the current odds of winning being smaller than 20 to 1), while in Shanghai the fee is more than one million JPY (9,800 US$). China has already partially introduced an emissions trading scheme for greenhouse gases, which was blocked in Japan by industry several years ago. These are examples of how China can only take compulsory steps to improve the situation—including population control and a ban on the construction of coal-fired power plants—that sacrifice the "freedom," "basic human rights," and "democracy" many advanced countries have come to enjoy. And it is China that is now actually taking these steps.

As a result, the ratio of coal consumption increases in the first half of 2014 became negative for the first time since the start of high economic growth following the open and reform policy. This meant the decrease of air pollutant emissions from coal in China. In fact, the emissions of sulfur dioxide, nitrogen dioxide, and carbon dioxide either stayed flat or indicated a declining trend. Many cities found declines in $PM_{2.5}$ concentration. For instance, in Beijing, 2010 and 2011 experienced the highest number of days of extreme pollution, and the annual average $PM_{2.5}$ concentrations were 117 and 98 µg/m³, respectively. However, the annual averages reduced to 82, 75, and 60 µg/m³ in 2015, 2016, and 2017, respectively (Liu et al. 2018).

According to Zhang et al. (2019), the estimated national population—weighted annual mean $PM_{2.5}$ concentrations decreased from 61.8 (95%CI: 53.3–70.0) to 42.0 µg/m³ (95% CI: 35.7–48.6) in 5 years, with dominant contributions from anthropogenic emission abatements (Figure 6.1)

The main reason for such decline was said to be the changes in air pollution measures, and the transition of industrial structures (Zhang et al. 2019). The most effective measure was the adoption of an extremely strict regulatory method to prohibit the use of coals, which could not be done except in a country of dictatorship like China. In that sense, there was not much influence of pressures from other countries, or atmospheric environmental control regimes in Asia.

If $PM_{2.5}$ concentration in China is halved, that will simply halve their contribution ratio to $PM_{2.5}$ concentration in Japan. Indeed, Japanese researchers published a paper titled "End of trans-boundary air pollution issue?" (Uno et al. 2017). According to that paper, the result of source-receptor analysis using the model developed by Uno et. al. showed about 12% decrease in the yearly average of $PM_{2.5}$ concentration in Fukuoka, for 20% decrease in $PM_{2.5}$ concentration in China.

In fact, $PM_{2.5}$ concentration in China did decrease around 20%, and Fukuoka City, which is in Kyusyu island and is close to the Chinese continents, observed about 10% decrease for the period of 2014–2016 (Figure 6.2). In other words, the predicted number was almost equal to the observed result. Therefore, Uno et al. (2017) predicted that, if emission decrease in China would continue at this rate, in a year or two, more observation points in Japan would observe

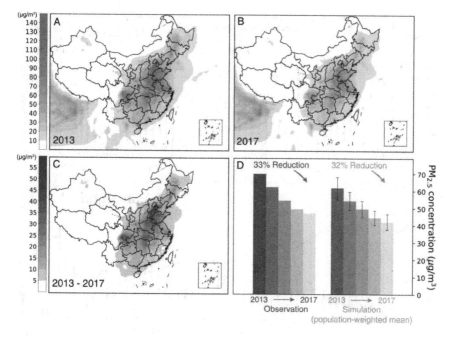

Figure 6.1 Variations in China's PM$_{2.5}$ concentrations from 2013 to 2017

Source: Zhang et al. (2019)

Note: (A and B) Distributions of annual mean PM$_{2.5}$ concentrations in China in 2013 and 2017. (C) Reductions of annual mean PM$_{2.5}$ concentrations between 2013 and 2017 (positive value). (D) Observed national annual mean PM$_{2.5}$ concentrations calculated based on samples from continuously operated sites and simulated national annual population-weighted mean PM$_{2.5}$ concentrations. Error bars: uncertainty ranges (95% CI) of model estimates

an annual average of PM$_{2.5}$ concentration meeting the standards, bringing rapid improvement in the issue of transboundary high-concentration PM$_{2.5}$ in Japan.

Following Uno et al. (2017), Uno et al. (2020) confirmed that a rapid decrease in PM$_{2.5}$ concentrations in China has been observed in response to the enactment of strong emission control policies. From 2012 to 2017, total emissions of SO$_2$ and NO$_X$ from China decreased by approximately 63% and 24%, respectively. Simultaneously, decreases in the PM$_{2.5}$ concentration in Japan have been observed since 2014, and the proportion of stations that satisfy the PM$_{2.5}$ environmental standard (daily, 35 µg/m³; annual average, 15 µg/m³) increased from 37.8% in fiscal year (FY) 2014 (April 2014 to March 2015) to 89.9% in FY 2017.

6.5.5   *Public support for the export of the technologies of coal-fired power generation*

Three countries of Japan, China, and Korea have been receiving strong criticisms from the international community for exporting their own technologies of coal-fired power generation, using their public funds. For example, the Japanese

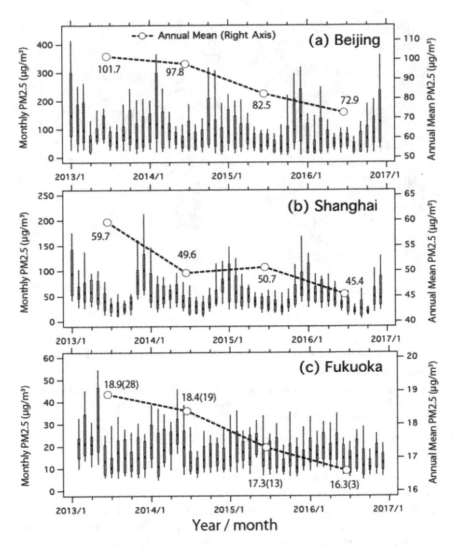

Figure 6.2 Monthly mean observed PM$_{2.5}$ concentration (box plot) for (a) Beijing, (b) Shanghai, and (c) Fukuoka between 2013 and 2016

Source: Uno et al. (2017)

Note: Open circle with dashed line shows the annual mean concentration (right axis)

government has supported the construction of coal-fired power plants around the world through public financial institutions such as the Japan Bank for International Cooperation (JBIC), Nippon Export and Investment Insurance (NEXI), and the Japan International Cooperation Agency (JICA). The Japanese government claims that the so-called "clean coal technology" will help reduce carbon dioxide emissions; however, this argument has been criticized heavily by many NGOs as well as researchers. Korea and China have also accelerated the export of the coal-fired power plant with their public fund.

This is one of the reasons why these countries are ranked very low by the Climate Change Performance Index, which is made by an international think-tank to evaluate the performance of the main emitting countries (Burck et al. 2020). For example, in 2020, Japan, China, and Korea were ranked No. 48, No. 37, and No. 55, respectively, out of 58 main CO_2 emitting countries.

It goes without saying that the export of coal-fired power plant will increase the local concentration of air pollutant such as $PM_{2.5}$, which contradicts with the air pollution mitigation objective of these countries.

6.6 Conclusion and discussion

It is not easy to institute an international regime for the atmospheric environment. This is because the measures adopted by such a regime will be likely to raise conflicts of interest. In many instances, countries tend to criticize other countries and try to shift responsibilities to others. In order to develop some kind of an agreement under such a situation, one must develop the infrastructure for science as well as for politics. Certainly, it is not simple work to establish scientific infrastructure that can provide sufficient objectivity.

In East Asia, countries have continued discussion for a long time without a consensus on the source-receptor relationship of each country. In 2020, however, the situation seems to be slightly changing with the publication of LTP model studies. At least, hopefully, from now on, regional countries may advance negotiation and discussion on such regime, based on the agreed result of quantitative analysis.

This will not result in the immediate agreeing on a regime for atmospheric environment. To build a regime, it is necessary to: 1) expansion and manifestation of damages; 2) (a certain degree of) consensus on the contribution ratios of pollution; 3) seriousness of national leaders in mitigating tensions; and 4) issue linkages (links with other energy and economic problems). In today's East Asia, unfortunately, most of the conditions above are not fulfilled.

On the other hand, there can be some arguments saying that the EU-like regime obligating emissions reduction is not necessary in Asia, as it is possible to implement drastic and mandatory measures in China. Moreover, it can be said that there seem to be no capabilities or ample political wills in East Asian countries to develop such an international framework.

Still, air pollution damages have not disappeared in the world and in Asia. The premature death toll of $PM_{2.5}$ is several million a year. $PM_{2.5}$ concentration in Asia is far greater than the WHO-recommended level of $10\mu g/m^3$ per year. With the rising voice calling for global warming measures, the pressures to reduce coal-fired power generation will continue to mount.

Recently, the world is focusing on the argument that the policy to promote energy transition, like Green New Deal, will bring positive effects on economy.

In the past, air pollution measures or global warming mitigation measures were thought to have adverse effects on economy. Today, widely spreading thought is that the expansion of green investment is the only way to realize economic growth and job creation, because of the declining cost of power generation using renewable energies, etc. Actually, jobs in the renewable energy field are rapidly expanding in the world, contributing to economic growth to a certain extent. Therefore, each country has an economic incentive to advance energy transition now.

Furthermore, technology cooperation is taking a different form nowadays. Technological cooperation used to be from developed countries to developing countries. Whether in Europe or Asia, technological cooperation tended to be discussed from the viewpoints of technology providers. Yet, the technological level of China is advancing rapidly. Today, it is no longer the problem of technological cooperation to China, but the technological hegemony of China (e.g., photovoltaic cell and nuclear energy) becomes a much bigger political problem for other countries.

It can be said that these three countries now recognize the criticism against their climate policy mentioned above and are trying to change the climate policy. On top of that, recently, the Korean government published the "Korean New deal" which is a quite ambitious policy package for green recovery after the COVID-19 pandemic. President Xi of China also declared that China will be carbon neutral in 2060 at the UN council in September 2020. On the contrary, recently, the Japanese government has introduced so-called "capacity mechanism" which will benefit keeping the coal-fired power plants.

As mentioned above, there are many diverse aspects in Asian countries and the situation is changing dynamically. Therefore, the environmental regime needs to be quite different from that of European countries. At least, it seems that LRTAP-type paradigm will not work in East Asia anymore. It is also true that each country has a lot of things to do domestically. Overseas investment in coal-fired power plants using public money should be stopped immediately as well. Without such actions, it is self-contradictory to talk about the importance of environmental problems or to try to build a regional or national regime to control atmospheric environment.

Notes

1 The first instance of addressing acid rain problem as a policy task in Japan was back in 1982 when Japan Environment Agency (EA) formed "the Acid Rain Measures Study Group." Later, in 1985, EA together with Forestry Agency conducted the "survey on actual situation of acidic depositions and deforestation of cedars in Kanto Area." In February of 1993, National Environmental Research Institute summarized the analytical data suggesting that they "demonstrated that particles causing acid rain were coming from the Continental Asia," since the concentration of sulfates and nitrates, which would cause acid rain, increased when there was a west wind over Japan Sea near Shimane coastline. In July of 1994, EA made the first official acknowledgement of the effects of nitrogen oxides from continental Asia in their report on the result of "acid rain measures survey" and expressed their concern for further increase in their effects for the future. Later

reports, such as the press release on the third acid rain measure survey by EA on March 19, 1999, seemed to indicate the decrease in the concerns of damage expansion and cause-effect relationship. Japan Environment Agency (EA) became Ministry of Environment Japan in 2001.

2 RAINS (Regional Air Pollution Information and Simulation Model) was developed by the IIASA (International Institute of Applied Systems Analysis) in Austria to consider the burden sharing among the countries.

References

Amann, Markus, Hordijk Leen, Klaassen Ger, Schopp Wolfgang, and Sorensen Lene (1992) Economic Restructuring in Eastern Europe and Acid Rain Abatement Strategies, *Energy Policy*, December, pp. 1186–1197.

Bjorkbom, L. (1999) Negotiations Over Transboundary Air Pollution: The Case of Europe, *International Negotiation*, Vol. 4, pp. 389–410.

Burck, J., et al. (2020) *Climate Change Performance Index 2020*. www.climate-change-performance-index.org/the-climate-change-performance-index-2020

EEA (1997) *Air Pollution in Europe 1997*, Executive Summary, European Environment Agency.

Ichikawa, Yoichi (1998) Long Distance Transportation of the Acidic Materials, *Journal of Atmospheric Environment*, Vol. 33, No. 2, pp. A9–A18.

IEA (2020) *Global Energy Review 2020*. www.iea.org/reports/global-energy-review-2020/coal

Ikeda, Kouhei, et al. (2014) Source Region Attribution of $PM_{2.5}$ Mass Concentrations over Japan, *GEO Chemical Journal*, Vol. 49, No. 2, pp. 185–194, 2015. https://doi.org/10.2343/geochemj.2.0344

Koyanagi, Hideaki (2019) *Debate on the Trans-boundary Air Pollution between China and Korea*, Global Net, INSIDE CHINA, February 19, 2019 (in Japanese).

Levy, Marc A. (1993) European Acid Rain: The Power of Tote-Board Diplomacy, in P. M. Haas, R. O. Keohane and M. A. Levy, eds. *Institution for the Earth*, Cambridge, MA: MIT Press.

Liu, Lun, Silva Elisabete A., and Liu Jianzheng (2018) A Decade of Battle against $PM_{2.5}$ in Beijing, *Environment and Planning A: Economy and Space*, Vol. 50, No. 8, pp. 1549–1552. https://doi.org/10.1177/0308518X18766633

Ministry of Environment Japan (2019) *Project on the Long-distance Transport of Air Pollutants in Northeast Asia*, Phase 4 (2013~2017) Summary Report (in Japanese).

Nagashima, T., et al. (2010) The Relative Importance of Various Source Regions on East Asian Surface Ozone, *Atmospheric Chemistry and Physics*, Vol. 10, pp. 11305–11322.

Uno, I., Wang, Z., Itahashi, S., et al. (2020) Paradigm Shift in Aerosol Chemical Composition Over Regions Downwind of China, *Scientific Reports*, Vol. 10, Article No. 6450. https://doi.org/10.1038/s41598-020-63592

Uno, Itsushi, et al. (2017) End of Trans-boundary Air Pollution Issue?, *Journal of Japan Society for Atmospheric Environment*, Vol. 52, No. 6, p. 177–184 (in Japanese). www.jstage.jst.go.jp/article/taiki/52/6/52_177/_article/-char/ja/. https://doi.org/10.11298/taiki.52.177

Wettestad, Jorgen (1996) *Acid Lessons? Assessing and Explaining LRTAP Implementation and Effectiveness*, IIASA Working Paper, WP-96-18.

Xuan, J., Soklik, I. N., Hao, J., Guo, F., Mao, H., and Yang, G. (2004) Identification and Characterization of Sources of Atmospheric Mineral Dust in East Asia, *Atmospheric Environment*, Vol. 38, pp. 6239–6252.

Yonemoto, Shohei (1994) *What is the Global Environmental Issue?*, Iwanami Shoten (in Japanese).

Zhang, Qiang, et al. (2019, December 3) Drivers of Improved $PM_{2.5}$ Air Quality in China from 2013 to 2017, *PNAS*, Vol. 116, No. 49, pp. 24463–24469. https://doi.org/10.1073/pnas.1907956116

Part III
National air pollution battles

7 Air quality injustice in Taiwan*

Just transition as the next chapter of environmental governance in post-developmental states

Kuei-Tien Chou and David Walther

7.1 Introduction

A scan of development research trends over the last 30 years has found that public health, environmental science, and risk governance have gradually become major areas of development research that have received increasing attention since 2003 (Han et al., 2014). As a former developmental state, Taiwan has adopted a strategy of playing catch up by making use of the late-mover advantage and has prioritized economic development as its national development policy. In the 1980s, developmental states started shifting from a model of bureaucratic capitalism to neo-liberal crony capitalism. Some scholars have referred to this as neoliberal developmentalism or the neoliberal-developmental states (Liow, 2012; McNeish, 2008; Heo, 2015; Féliz, 2012).[1] Studies have pointed to neoliberalism being the main roadblock in the ability of post-developmental states to undergo a paradigm shift to a different model (Chang, 2012; Kiiza, 2012; Ban, 2013; Hochstetler and Montero, 2013).

One of the major dilemmas of post-developmental states is the conflict between environment conservation, economic growth, and just issues. Their growth strategy continues to rely on underestimating the external costs of the environment and health, in order for businesses to amass profits to sustain growth. This is the "system of sacrifice" among the developmental countries predicated upon predatory economics and accumulation by dispossession (Chou and Walther, 2016). Post-developmental states have therefore veered toward becoming "neoliberal-developmental states" over the last 30 years putting them in opposition with sustainable development—this is in comparison to the sustainable development efforts undertaken by European industrialized countries (although the latter have not succeeded in ridding neoliberalism as well). Regarding air pollution governance, Taiwan lacks the drive to transition to a new paradigm, while policies that could more actively address the issue have been delayed or put on the back burner (Walther and Chou, 2018; Chou et al., 2018; Tu et al., 2017). In addition to dealing with the conflict between environmental

* We would like to thank Roy Ngerng for his decent proofreading of the chapter.

DOI: 10.4324/9781003211747-10

sustainability, economic growth, and just issues, the next stage of environmental governance and sustainable transition among post-developmental states is to embed the concept of just transition and change the paradigm of current environmental governance.

In Section 7.2, the chapter will analyze the development of environmental governance within post-developmental states, by first exploring the relationship between Developmentalism and the Post-Developmental states, and understanding how globalized neoliberalism has become the main obstacle to environmental governance within Post-Developmental states. Section 7.2 will also detail the predicament confronting Just Transition in the face of neoliberal dominance and introduces a reference framework that can be used to analyze the development of just transition in Taiwan. In Section 7.3, the research methods, as well as data processing and analysis approaches, and the research limitations will be detailed. In Section 7.4, data on the socioeconomic situation in Taiwan's counties and cities is compared with PM2.5 concentration data to illustrate the unequal distribution of air pollution in Taiwan, as well as to highlight the severity of the problem in areas with lower incomes and higher agricultural production. The section also details how in spite of improvements to air quality in Taiwan in recent years, the situation is being reversed. In Section 7.5, we detail how although Taiwan's Environmental Protection Administration (EPA) has commissioned several air quality studies covering the whole of Taiwan, the development of local policies have however not been evidence-based, in how policies have not been differentiated according to the pollution levels in different regions, thereby leading to the inconsistent allocation of pollution risks. Using the analysis framework of just transition by Heffron and McCauley (2018), we therefore conclude that just transition has not received much attention in the environmental governance of Taiwan's air quality policy. This study advocates for the use of the analysis framework as a guide to implementing just transition for the improvement of Taiwan's air quality policy.

7.2 Environmental governance in the post-developmental states

7.2.1 *Neoliberal developmentalism as an obstacle to sustainable development*

Although there are different genres in the studies of developmental states, these states generally adopt similar strategies, in terms of deliberately distorting existing price relativities and other market elements via selective tariffs and subsidies and biased access to finance, in order to interfere with the pace and direction of capital accumulation. Meanwhile, the authoritarian character of the state ensured that the competing interests of specific groups and sectors become subordinated to national development goals, which are largely dependent on industrialization and technological needs anyway (Radice, 2008). However, from the 1990s, the countries first categorized as developmental states, Taiwan, South Korea, and Japan, were observed to have shifted toward neoliberalism, as part of a global neoliberal wave that spread throughout the world. In South Korea, the use of neoliberal political rationality and governmentality to pursue its international, political, and socioeconomic goals has led to the development of "neoliberal developmentalism" which is under the global trend of neoliberalism (Lee, 2008; Suzuki et al., 2010; Gao, 2013; Hashimoto, 2014). Nonetheless, regardless of whether Taiwan is studied under the paradigm of a developmental state or globalized neoliberalism, an in-depth critique of developmentalism is in order, if not Taiwan will continue to face barriers transiting into a post-developmental state with sound ecological and environmental governance (Hsia, 2015).

The common theme that runs through these post-developmental states is their historical economic development. The economic models of Japan, South Korea, and Taiwan are driven by fossil-based energy which requires high energy consumption and produces high carbon emissions. Even as these countries are moving toward sustainable development goals, such as by decoupling carbon emissions from GDP, increasing the proportion of renewable energy, or improving the quality of the environment and health, these post-developmental states have however yet to achieve the same level of sustainability as other advanced countries.

The rapid liberalization of Taiwan in the 1980s should be seen in the context of global trade liberalization, the decline of manufacturing in the USA, and the increasing international division of labor, which has led to challenges to the state-led economic model, and strengthened calls for full liberalization. Such an economic model of developmentalism resulting from "regime dynamics" has however led to developmental path dependence (Polanyi, 1944), which has hindered the ability of these developmental states to develop based on the principles of environmental sustainability, fairness, and justice (Cheng, 2014; Hsia, 2015; Chou and Walther, 2016). With the advent of neoliberalism in the 1980s, its developmental ideology has led to deregulation, privatization, and marketization, resulting in industries becoming even more energy and water intensive, and polluting, thereby putting in the back burner Taiwan's transition toward sustainability (Chou, 2017). On the other hand, the liberalization and democratization in Taiwan and South Korea over the past 40 years has also led to progress in terms of social equality, social welfare and social reforms, as well as environmental sustainability (Cheng and Li, 2010; Chen, 2005; Ramesh, 2003; Wong, 2006). Although Taiwan's liberalization and democratization after 38 years of martial law have enabled it to reestablish its regulatory capabilities to push for several environmental sustainability policies, the ideology and regime of developmentalism have been firmly established. In other words, throughout its time as a developmental and post-developmental state, Taiwan has failed to develop its capabilities for sustainable environmental governance (Chu, 2007; Weiss and Hobson, 1995; Chou and Walther, 2016).[2]

However, facing pressure from international conventions with regard to sustainable development and carbon reduction, in order for Taiwan's economic system to be transformed into a sustainable one, this would require urgent policy interventions to limit the carbon emissions and electricity consumption of the petrochemical industry, as well as increase the proportion of natural gas and renewable energy generation (Chou et al., 2019). However, the implementation of such a strategy has encountered roadblocks, one of the fundamental reasons being that Taiwan has become overly reliant on cheap electricity and water, and high carbon emissions, in its bid to maintain competitiveness in the neoliberal global economic system. Although the electronics industry is the major contributor to Taiwan's GDP, the petrochemical and energy industries are significant contributors as well, but while their industrial transformation is required, requires profound changes to Taiwan's economic and social development (Chou et al., 2019).

7.2.2   *Air quality just transition as a means to confront neoliberal dominance*

In this article, just transition can be defined in the simplest terms as "appropriate transition based on just principles." According to Heyen et al. (2020), the term "Just Transition" emerged in the late 1970s when the US Oil, Chemical, and Atomic Workers Union sought support for workers whose jobs were threatened by environmental regulation (Gambhir et al., 2018). The definition of Just Transition has not yet been unified in different disciplines and applications, but the description given by Bennett et al. (2019) is the most compatible and most usable:

> *Deciding what constitutes a fair or just distribution is complicated by underlying philosophies and 'equity criteria' (i.e., utility, equality, proportionality, needs, merit, and rights [. . .]) that are often implicit in environmental and sustainability decision-making processes. Moreover, the interlinkages between ecological sustainability and social justice are manifold.*

When just transition is transformed from a concept to a specific policy goal, the pragmatically defining just transition policy goals (EnvP) from the directorate-general for environment, European Commission (Franssen and Holemans, 2020), it is:

> *(1) EnvP reduce inequalities in the distribution of environmental [bads & goods] and with regard to social inclusion. (2) EnvP themselves do not disproportionally burden vulnerable/underprivileged households and ensure that financial (saving) opportunities are also available to them. (3) EnvP positively affect quality & quantity of employment and, together with structural policy, they also open up perspectives for workers & regions affected by the transition.*

Our current system generates massive economic and environmental inequalities—both within countries and between them. Environmental justice means attacking these inequalities at their root, ensuring that no community bears the excess burden in the climate and environment emergency and that all communities gain together from our transition out of it. Many decision-makers, companies, and researchers often fail to incorporate "justice" when designing strategies to improve environmental quality and achieve sustainability, but this results in environmental sustainability being overlooked thereby exacerbating injustice, which makes it even more pertinent to include the concept and practice of just transition into transition planning (Heffron and McCauley, 2018). Therefore, a society moving toward improving air quality should fully consider the process and results of a just transition.

The central problem for CEE justice research, according to Heffron and McCauley (2018), is the dominance of neoliberalism. While neoclassical economics should in theory allow for competitive markets, the neoliberal agendas have led instead to the "opposite result," they contended. Traditional economics has therefore not delivered "just" positive outcomes for society, nor have traditional CEE science and engineer scholars. If anything has been achieved, it is that social economics has allowed for social equality to be discussed in the field of environmental research, but even so, the power of discourse in the research between social justice and the environment still lies largely within the domain of traditional economics (Heffron and McCauley, 2018). Within traditional environmental research, neoliberal environmental economics continues to dictate economic policymaking and dominate the discussion of environmental policy.

Neoliberalism has allowed companies to profit through accumulation by dispossession; by suppressing taxation and labor protection to reduce business costs (Harvey, 2007), while politicians and capitalists work together to achieve financialization via privatization, crisis management, and the manipulation of national mechanisms, such as of the governance structures and systems over the environment, land, water, and labor, or in other words, via a form of rent-seeking exploitation (Harvey, 2005). In the end, the pollution and risk producers reap most of the benefits, while the risks are transferred to local residents and left to them to manage (Chou and Walther, 2016). For example, the PM2.5 standard Taiwan set in 2012 is behind the USA by 15 years and 7 years behind the World Health Organization (WHO)'s recommendation, and while the policy target was for Taiwan to reach the 15µg/m3 standard by 2020, it was however postponed to 2023 in 2020. Taiwan's EPA has long heeded the 'recommendations' made by the Ministry of Economic Affairs and the Chinese National Federation of Industries (CNFI), based on the fear that not doing so and imposing additional regulations will impact the industries and result in economic losses (Walther and Chou, 2018). Therefore, the key problem Post-Developmental states face in their environmental governance is the issue of how to overcome developmentalism, and move toward sustainability and justice.

The UN human right council has listed the "right to breathe clean air" as a human right (UN, 2019). From a global scale, it can be found that the adverse health effects of air pollution are highest in low-income and middle-income countries where exposures to both ambient and household air pollution are high. The UN report points out that air pollution disproportionately harms poor people and poor communities and poverty exacerbates the impacts of air pollution through lack of access to information, health care, and other resources. A considerable amount of research has shown that low-income and minor people have long borne an unequal burden of environmental health threats in their neighborhoods compared to the general population (Morello-Frosch and Jesdale, 2006; Mohai et al., 2009) and a number of environmental inequalities related to air pollution have also been raised (Hajat et al., 2015; Banzhaf et al., 2019; Tessum et al.,2021). However, most of these studies talk about "air pollution" inequalities rather than the potential injustices of improving air pollution. As discussed in the EEA report (EEA, 2018), while the ban on the most polluting fuels can improve the air quality and public health, it is also associated with equity issues related to some negative economic equity consequences such as the affordability of fuel and the ability of poorer households to cope with cold temperatures.

Scholars in Taiwan have discussed Taiwan's unjust development model in their research. Jobin (2020) pointed to how the agriculture sector in the relatively poor county of Yunlin is facing the injustice of environmental and public health damage resulting from one of the largest chemical companies in the world, mainly due to the political corruption, patronage system, and ambiguous means by which compensation is given to mediate disputes under Formosa Plastics's "good neighbor" policy in local areas. Shih and Tu (2017) also found that policy decisions based on information from singular sources, particularly from polluters, both erode the legitimacy of the government's policy decisions and reduce their ability to provide adequate oversight. Several studies have also highlighted the knowledge imbalance between environmental regulation and scientific assessments (Chou et al., 2018). In the end, the way scientific and technological knowledge is being manipulated by large chemical companies and the government has hampered the accident handling procedures within these companies, resulting in the failure of restorative justice (Tu and Shih, 2019).

These studies, therefore, illustrate the consequences that Taiwan is facing due to the policies of deregulation, privatization, and marketization it has undertaken in its laissez-faire capitalism strategy. As Taiwan moved from being a developmental state to neoliberalism, it simply did not establish a governance tradition of environmental and social justice, and these studies therefore confirmed that Taiwan's move toward neoliberalism has increased the vulnerability of labor and the environment to risks and that the injustice caused by such structural exploitation and risk transfer has instead become part of the logic of Taiwan's economic development (Chou and Walther, 2016). Taiwan's continued disregard for environmental sustainability and social justice over the past 40 years has therefore exacerbated the environmental dispossession and unjust risk distribution in Taiwan.

A just transition should be a societal goal, and the design and implementation of transitions such as the promotion of good air quality, and the achievement of a low carbon society and the sustainable development goals (SDGs) need to abide by the same principles of a just transition, but any such action should include measurable outcomes. The assessment of justice cannot however be quantitatively measured but would require qualitative research for the assessment of any impacts to the law and/or policy, or to collate reports of whether previous

Table 7.1 Reference framework for just transition

Conceptual Dimension	Assessment Component
Justice	Distributional justice
	Procedural justice
	Restorative justice
Universal	Recognition justice
	Cosmopolitanism justice
Space	Local, national, or international
Time	Transition timelines
	Start and end time
	Duration

unjust decisions were overturned or improved (Heffron and McCauley, 2018). Heffron and McCauley (2018) have proposed a reference framework based on the study of legal geography that can be adopted to guide society toward just transition (Table 7.1). Importantly, public acceptance and understanding are crucial as to whether the public will take action and support such a transition.

Under this framework, just transition is assessed in terms of distributional, procedural, and restorative justice (Heffron and McCauley, 2018). Distributional justice refers to how the distribution of benefits and risks in a transition should be equal; procedural justice emphasizes the need for transition processes to conform to principles of justice, such as in terms of democratic participation and interest avoidance; and restorative justice is especially pertinent when faced with unequal treatment during transition processes, such as for matters like compensation. Universality under the framework refers to how the recognition of justice and cosmopolitanism justice should be assessed, and whether a compromise can be made, with the former (justice recognition) focused on the need for intersubjective recognition among stakeholders when seeking just action, and the latter (justice of cosmopolitanism) which transcends specific stakeholders, tribes, and nations, and is aimed at upholding universal human values. Using this framework, we will analyze Taiwan's air quality policies and evaluate the injustices of environmental transition, by way of air pollution health risk assessments which have been typically conducted using population-based and cohort study designs, the former of which enables evaluation over the justice of spatial distribution, and the latter over the justice of generational (time) distribution.

7.3 Research method and limitations

This study aims to use the just transition framework to study Taiwan's current air quality policies and explore the options for adopting just transition in the next stage of Taiwan's environmental governance as a post-developmental state, by analyzing the distribution and reduction of PM2.5 air pollution, and comparing them with socioeconomic data, and the evolution of Taiwan's air pollution situation and policies.

Based on the theoretical context of developmentalism in the developmental states, this research studies the inequality of air pollution and the just transition that must be undertaken in order to improve air quality, as well as develop new insights on environmental governance in post-developmental states like Taiwan. Statistical analysis is performed using PM2.5 pollution and socioeconomic data and tested on Heffron and McCauley (2018)'s framework to assess whether Taiwan's air quality and policies can meet just requirements.

This research uses air quality and socioeconomic data for analysis, rather than by estimating policy effects using air quality models. While the extent to which inequality can be tackled by just transition cannot thus be estimated by the model to provide decision-makers with outcome metrics to reduce regional differences, this study however elucidates the phenomenon of inequality so as to reinforce the need to face squarely the just transition issues in Taiwan. In addition, this study highlights that just transition is a transitional problem that Taiwan must address early on, if it is to follow on the pathway of sustainable transition that other wealthy industrialized countries have embarked on with more effectiveness. Suggestions on how just transition functions can be put into practice will also be briefly touched on.

7.4   Air quality governance and issues of justice in Taiwan

7.4.1   Air quality improvement in Taiwan

7.4.1.1   The PM2.5 situation in Taiwan

In 2015, outdoor PM2.5 pollution ranks fifth among the causes of death collaborators in the Global Burden of Disease study, causing 4.2 million deaths worldwide each year, while 600,000 children die from air pollution each year (Cohen et al., 2017: 1910–1917). Only less than 8% of the population in the Asia-Pacific region live in environments where the air quality does not pose a significant risk to their health (UNEP, 2019). The air pollution in Taiwan is among the worst in the world. Figure 7.1 shows that in terms of PM2.5 concentrations, which are indicative of air quality, only 5 of Taiwan's 31 manual sites (including two in national parks) meet the WHO annual average of 10 μg/m3 in 2020, and 13 sites did not meet the EPA's standard of 15 μg/m3 (TEPA, 2021). The average annual PM2.5 concentration in Taiwan was as high as 24 μg/m3 in 2013, but it has since been reduced to 14.1 μg/m3 in 2020, which is still higher than the WHO recommendation of 10 μg/m3, as a result of the national anti-air pollution campaign and policy mainstreaming.

Based on the EPA air pollutant concentration data, the annual average concentration of major pollutants listed on Taiwan's Air Quality Index (AQI) has met the air quality standard, except for PM2.5. In addition, as PM2.5 is derived

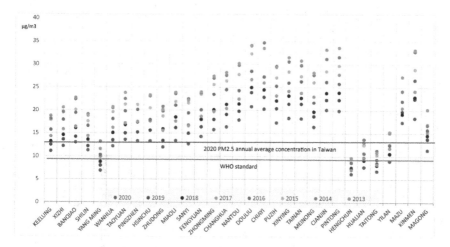

Figure 7.1 PM2.5 concentration in Taiwan (2013–2020)
Source: Taiwan's Environmental Protection Administration (TEPA)

from the main air pollutants nitrogen oxide (NOx), sulfur oxide (Sox), and volatile organic compounds (VOCs), which then react in the presence of sunlight to produce ozone (O3), another major air pollutant, PM2.5 concentration therefore not only reflects the concentration of particulate pollutants in the air but can also be used as an index to measure air quality to a certain degree. Therefore, this article focuses on Taiwan's PM2.5 pollution situation as the basis to analyze the air quality governance and just transition issues.

The Global Burden of Disease (GBD) study which uses a comparative risk assessment framework to calculate the disease burden in Taiwan, found that in 2019, particulate air pollutants (including PM10 and PM2.5) accounted for 1,161.51 Disability Adjusted Life Years (DALYs) per 100,000 people. In comparison, the DALYs in other Asian industrialized countries like Japan, South Korea, and Singapore were 602.16 DALYs, 953.77 DALYs, and 616.85 DALYs, respectively, and when compared with the European industrialized countries, it was 463.35 DALYs in the United Kingdom (UK), 669.84 DALYs in Germany, 557.3 DALYs in the Netherlands and 388.44 DALYs in France. The DALYs was 373.15 years in the USA, 2,310.52 DALYs in China, and 633.22 years in the Organization for Economic Co-operation and Development (OECD). It is apparent that the disease burden in Taiwan due to particulate air pollutants was relatively higher than in these other countries, which emphasizes the challenges that Taiwan's air quality management faces.

Figure 7.2 shows that there were two waves of improvement in 2014–2017 and 2018–2019. Figure 7.2 also illustrates how divergent the improvement in air quality is across the country is, where the situation is even deteriorating in some areas. In Taiwan, although PM2.5 pollution has heavily impacted people's

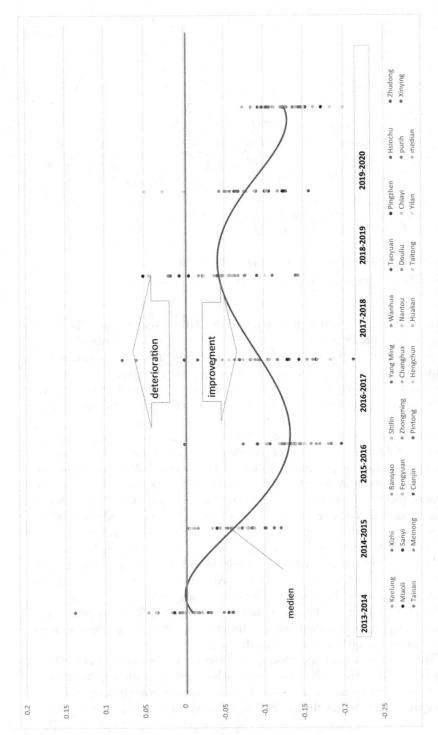

Figure 7.2 PM2.5 improvement in Taiwan (including the median change) (2013–2020)

Source: TEPA

health, and the academic community which began researching on this in 2000 had called on the government to manage and reduce PM2.5 levels, the setting of the PM2.5 standard to 15 µg/m3 was however postponed until 2012. The target was later revised upward to 18 µg/m3 in 2015, but even then will only be implemented in 2017. And when 2020 came, the government again postponed the deadline to achieve the target of 15 µg/m3 to 2023. The repeated delays raised questions about the government's air quality governance, and whether the EPA has encountered a bottleneck in its air quality improvement program.

7.4.2 The pattern of uneven air pollution distribution in Taiwan

From the perspective of governance, Taiwan's air quality governance needs to enter into a new era. Existing regulatory tools are facing "declining (marginal) benefits, while the effectiveness of policy measures is declining as well." In short, Taiwan's environmental governance is facing roadblocks, and without new concepts and tools, improvements to Taiwan's air quality will be increasingly challenging.

Figure 7.2 shows that there has been air quality improvement over the past 7 years, though it varies from region to region. After Taiwan's eastern region, the northern region has the next highest chance of reducing its PM2.5 concentration to 15µg/m3, but change has been slow-moving and levels still far exceed WHO's standard. Taiwan's air quality also deteriorates as we move southward, with Kaohsiung having the worst air quality on the island (not surprising given the presence of heavy industries such as the petrochemicals, steel, and thermal power generation industries), followed by Yunlin, where the location of one of the world's largest petrochemical companies, the Formosa Petrochemical Corporation Company, is located.

In order to have a deeper understanding of the unequal distribution of air pollution in Taiwan, the study extracted 2015–2020 data from 74 Federal Equivalent Method (FEM) monitors and divided them into five quadrants for analysis.[3] The results showed that the higher the concentration, the lower the air quality improvement: as Figure 7.3 shows, the first quintile has an average PM2.5 concentration of 23.3 µg/m3, but the air quality improvement rate is only 12.21%, while the last quintile has a concentration of only 9.7 µg/m3 but has the highest improvement rate of 26%. The average improvement rate is 21.3%.

Figure 7.4 uses the same statistical method for analysis as Figure 7.4 but sets the base year for comparison at 2015. It can be seen from the chart that the most heavily polluted areas in 2015 achieved an improvement rate of 22.44%. This signifies that in 2015, while some heavily polluted areas showed significant improvement, other areas could not replicate similar success, resulting in a low rate of improvement as seen in Figure 7.4. In other words, as of 2019, the most polluted areas have seen the least improvement over the past years, and these areas are thus the most concerning for Taiwan.

Table 7.2 similarly analyzed the PM2.5 improvement rate from 2015 to 2019 by quadrant, to identify the intersection between the two quintiles with the

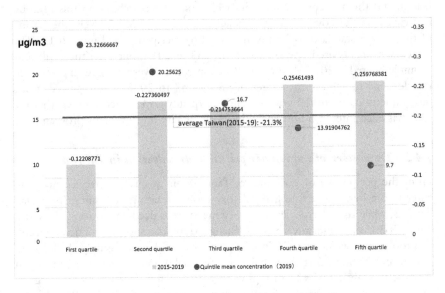

Figure 7.3 Quintile analysis of PM2.5 concentration and improvement rate
Source: TEPA's FEM Monitoring Site[4]

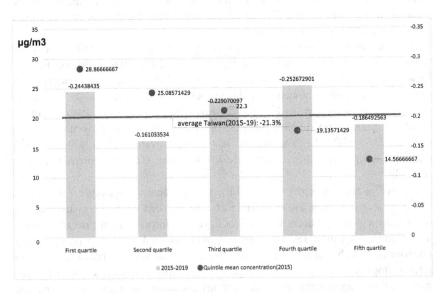

Figure 7.4 Quintile analysis of PM2.5 concentration and improvement rate (using 2015 for base year)

Source: TEPA's FEM Monitoring Site

Table 7.2 The two quintiles with the least improvement in PM2.5 (2015–2019)

city or county	site name	2013	2014	2015	2016	2017	2018	2019	2020	2015–2019
Kaohsiung City	Nanzi	32.70	31.80	23.4	25.7	29.4	27.7	24.7	19.7	5.56%
Kaohsiung City	Qiaotou			24.7	22.6	31	32.2	24.9	20.1	0.81%
Taoyuan City	Zhongli			18.4	19.7	17	15.6	18.3	14.7	-0.54%
Hsinchu County	Hukou	23.21	23.86	16.1	20.6	18.8	18.0	15.8	13.4	-1.86%
Tainan City	Annan	30.51	29.49	24.5	26.5	24.3	23.1	23.2	19.4	-5.31%
Pingtung County	Hengchun	9.60	10.17	9.8	6.3	10.1	8.8	9.1	7.2	-7.14%
Tainan City	Tainan	30.10	28.22	24	27	25.1	23.1	22.2	18.5	-7.50%
Kaohsiung City	Qianzhen			22.6	23.1	29.5	23.4	20.7	19.0	-8.41%
Kaohsiung City	Renwu	32.42	32.47	26.5	27	27.9	24.3	23.9	21.3	-9.81%
Chiayi County	Xingang	28.47	29.28	25.2	25.4	24.5	24.9	22.7	18.2	-9.92%
Kaohsiung City	Qianjin	32.72	30.42	25.9	28.7	26.7	25.0	22.9	20.1	-11.58%
New Taipei City	Tamsui	19.31	20.63	14.2	18.8	14.9	14.7	12.5	11.3	-11.97%
Kaohsiung City	Fengshan			25.30	26.4	27.7	25.0	22.2	19.7	-12.25%
New Taipei City	Sanchong			16.8	18	15.9	14.4	14.6	16.6	-13.10%
Taipei City	Datong			17.7	17.8	15	16.3	15.3	16.0	-13.56%
Taoyuan City	Guanyin			19.00	21.3	18.9	21.4	16.4	13.6	-13.68%
Pingtung County	Chaozhou	33.21	29.85	27.1	29	27.4	24.2	23.1	19.9	-14.76%
Pingtung County	Pingtung	33.62	29.91	26.6	27.7	27.6	25.2	22.6	19.4	-15.04%
Taitung County	Taitung	10.84	11.40	9.7	8.9	8.8	8.6	8.2	6.8	-15.46%
Kaohsiung City	Daliao	33.82	32.58	29.7	27.2	25.9	27.6	25.1	22.7	-15.49%
Tainan City	Xinying	30.40	29.99	25.6	26	25.7	26.5	21.6	17.6	-15.63%
Miaoli County	Miaoli	23.78	25.71	23	22.4	19.8	22.3	19.2	13.5	-16.52%
Kaohsiung City	Linyuan	33.32	29.79	28.1	25.5	21.6	22.7	23.3	20.9	-17.08%
Taichung City	Fengyuan	22.04	23.46	21.9	23.2	18.3	18.1	18.1	14.6	-17.35%
New Taipei City	Cailiao	21.43	25.43	16.5	16.7	16.4	15.6	13.6	13.0	-17.58%
Yunlin County	Douliu	33.56	34.42	29.7	26.7	26.5	24.5	24.2	19.9	-18.52%
Taipei City	Guting	21.13	19.69	17.5	16.7	17.1	14.0	14.2	11.9	-18.86%
Yunlin County	Taixi			24.7	24.8	25.7	21.9	19.9	16.1	-19.43%

Source: TEPA's FEM Monitoring Site

worst improvement rates (29 sites) and the areas in Taiwan with the worst average concentration in 2015 (reaching 20μg/m3) (47 monitors). A total of 17 intersections accounting for 61% of the total monitors were identified, of which three were in Tainan City (accounting for 75% of the monitors in the area); two in Pingtung County (accounting for 100% of the monitors, excluding those in the national park); eight in Kaohsiung City (accounting for 53% of monitors); one in Chiayi County (accounting for 50% of the monitors); two in Yunlin County (accounting for 50% of the monitors); and one in Taichung City (accounting for 20% of the monitors).

In summary, although many heavily polluted areas achieved significant improvements in 2015, 61% of them still fell into the two quintiles with the least improvement. It is worth noting that these areas with high pollution and low improvement are concentrated in Tainan County, Kaohsiung City, Pingtung County, Yunlin County, and Chiayi County.

7.4.3 *The regional injustice of air quality distribution in Taiwan*

Taiwan's air quality and improvement rates are unevenly distributed, as Figures 7.3 and 7.4 and Table 7.2 show. The question that needs to be further explored is the demographic makeup of the areas with high levels of pollution but yet lack improvement, and whether there are similarities or particular characteristics in these areas which could lead to issues of injustice.

In this study, the 19 counties and cities on the island were ranked according to their per capital disposable income, alongside which their PM2.5 concentration, production value of agriculture, and selected socioeconomic conditions (educational level, health indicators, and crude death rate) are listed in Table 7.3,[5] with the rows shaded according to a color scale indicated by light gray at the 10th percentile, dark gray at the 90th percentile. The comparison indicates the following:

1   Income: The top five counties and cities with the lowest incomes are high-polluting areas; with the 12th highest-income city of Tainan also highly polluting.
2   Production value of agriculture: the two sets of data illustrates that the areas where the production value of agriculture is high are also areas with low incomes and high levels of pollution.
3   Educational level: the two sets of data on education illustrates that education and income levels are correlated, and correspond roughly to pollution levels.
4   Healthcare indicators: there are no significant differences in the distribution of healthcare institutions and personnel, but the crude death rate is consistent with areas with low incomes, high levels of pollution, and higher agricultural value. As explained previously, high-polluting areas tend to see fewer air quality improvements, which accordingly means that the areas in Taiwan with lower income and educational levels, higher mortality, and greater agricultural production, are trapped in a cycle of air pollution.

Counties (2019)	per Capita Disposable Income (2019)	particulate matter 2.5 (PM2.5) (2019)	production value of agriculture, forestry, fishery and livestock farming industry (2019)	production value of agriculture (2019)	the educational structure of the employed people-Colleges and above	the educational structure of the public population aged over 15-Colleges and above	Average number of persons served per medical institution	Medical practitioners per 10,000 population	crude death rate
Taipei City	458,839	14	567,594	503,551	79	75	722	222.02	6.78
Hsinchu City	401,738	16	800,336	231,369	62	56	1,015	149.80	6.24
New Taipei City	371,156	13	7,971,670	3,177,603	53	46	1,192	97.78	6.26
Taoyuan City	357,432	16	11,799,493	6,858,021	49	44	1,395	124.65	5.91
Hsinchu County	355,984	15	6,572,298	3,975,606	53	45	1,382	94.90	6.74
Taichung City	352,633	18	31,090,924	27,050,250	52	48	803	153.83	6.31
Chiayi City	347,938	22	637,040	561,070	63	58	651	249.62	7.64
Kaohsiung City	331,657	22	36,042,243	14,666,159	51	46	910	153.58	7.87
Keelung City	328,890	12	4,764,427	49,474	52	43	1,206	117.00	8.52
Hualien County	322,574	9	9,209,094	6,889,468	45	39	1,113	182.93	10.03
Yilan County	311,283	11	13,050,409	6,853,735	44	37	1,294	135.06	8.78
Tainan City	305,444	22	57,329,110	29,314,553	46	42	962	146.40	8.22
Miaoli County	305,114	17	13,979,863	11,034,163	42	36	1,424	93.08	9.26
Taitung County	297,520	8	12,520,096	10,305,635	32	29	1,314	123.90	11.11
Yunlin County	287,962	23	79,202,144	38,320,584	37	32	1,341	101.75	10.56
Changhua County	285,842	20	60,960,140	27,946,438	41	36	1,199	120.68	8.29
Pingtung County	284,524	23	67,874,059	23,918,466	37	34	1,253	117.98	10.21
Chiayi County	271,154	21	42,250,732	23,887,634	33	30	1,857	110.41	11.02
Nantou County	270,085	21	31,583,284	26,639,545	40	36	1,154	105.93	9.92

10 percentile
30 percentile
50 percentile
70 percentile
90 percentile

Source: TEPA

The aforementioned regional inequality in the ineffectiveness of air quality governance has therefore deepened people's distrust in the government. However, because past regulations have predominantly improved air quality in affluent areas, this has created the impression that the government's air pollution management policy is unfair, thus making it difficult for the government to further promote its policy tools. Another major challenge to just transition in Taiwan is the prevalence of small and medium-sized enterprises in Taiwan which while playing an important role in the livelihoods of Taiwanese, are confronted with issues of low capital, even as some of them have capital comparable to large enterprises. However, this does not mean that one can turn a blind eye to the high levels of pollution generated by these small factories, especially so for the contractors, employees, and supply chain, as well as other stakeholders of these factories. Stakeholders including local residents, academics, and civil society are equally important as well. What is worrying however is that Taiwan has yet to set up a platform for multi-stakeholder cooperative governance based on mutual trust between government officials, the public, and industry. Taiwan's ability to undergo just transition in its air quality improvement strategy is, therefore, being severely tested by its economic model.

7.5   Insights for air quality governance based on just transition principles

As Taiwan's government air improvement policies are developed based on the Air Pollution Control Plan that is reviewed every four years, this study's analysis should first and foremost take an incisive look at the plan. According to the plan, the target for Taiwan's average PM2.5 concentration standard is set at the national level, but not for every city/county, but based on the National Ambient Air Quality Standard (NAAQS) adopted by the US Environmental Protection Agency (US EPA), it is a standard the federal government requires of states to achieve and is not a national-average target. In fact, the national average in the US is usually lower than the NAAQS, and it is required by states that their State Implementation Plans (SIPs) comply with the NAQQS, which therefore acts to standardize air quality policy goals across the states. The US EPA, therefore, plays the role of monitoring the air quality in the states and requires them to achieve the NAAQS.

Given the size of Taiwan, TEPA can consider modeling itself after a state in the USA, specifically the state of California, which has the most advanced air quality management system in the USA. California has adopted atmospheric sciences (specifically by studying the actual atmospheric flow conditions) to divide its 'air basins' into Air Quality Management Districts (AQMD), within which different policies and measures have been implemented in each district. Accordingly, the South Coast Air Quality Management District was formed to manage one of the worst air pollution problems in the Greater Los Angeles Area, based on the air basin areas. However, Taiwan's air quality management is not strategized according to its air basins, but via administrative regions, which results in the Hengchun Township being included in the Kaohsiung-Pingtung Air Quality Zone (Kao-Ping AQZ), the region with the worst air quality in Taiwan. This is even though Hengchun is a very clean township located within the National Park Reserve. Environmental non-governmental organizations (NGOs) and some scholars believe however that this instead results in the air quality data of the Kao-Ping AQZ being positively inflated. Therefore, it can be said that the logic of the air quality management zone is one of the reasons for the persistent air pollution and unequal air quality improvement in Taiwan.

Another important aspect is that, compared with the USA, Taiwan's evidence-based policymaking is less integrated and cogent. The documents consulted on during the formulation of the NAAQS in 2012 were a 2228-page integrated scientific assessment report, a 458-page policy assessment report, a 478-page impact assessment report, a 596-page health risk assessment report, a 474-page regulatory impact assessment, and other assessment reports, as well as assessments and feedback from expert committees, all of which are accessible in the public domain. EPA also solicited feedback from people from all walks of life and industries, to ensure that the policies developed are grounded in evidence. On the other hand, the scientific and policy assessment reports consulted by TEPA in the development of policies have not been made publicly available, and while researchers have a rough idea of the commissioned research that policies developed are based on, there is however no official scientific documentation of how TEPA translated such research and assessments into the regulatory standards and policies. TEPA often prides itself on adopting "evidence-based" management approaches in the development of air quality policies, when in actuality the governing of Taiwan's air quality management is based on existing administrative divisions, and thus lacks evidence basis.

This study will use the "just transition framework" proposed by Heffron and McCauley (2018) to analyze the issues with Taiwan's air quality management in order to provide insights. First, as mentioned in the previous section, TEPA does not place emphasis on *distributional justice* in its overall policy planning for air quality. While TEPA's policy adopts an air quality model that is based on a scientific assessment of the amounts of air pollutant concentrations to lower in order to reduce air pollution, and which also has the ability to simulate the regional differences in concentration and reduction, it does not, however, integrate the social and economic realities into its scientific assessment. As can be seen, the pollution in areas with low-income and low socioeconomic conditions, and which have relatively higher agricultural production, are very serious; however, Taiwan does not structure its policy differentiation according to pollution districts (as California has), but continues to use administrative regions for its policymaking. Therefore, TEPA needs to follow suit by setting differentiated standards and special measures for the highly polluted areas.

Procedural justice is important in introducing democratic supervision to ensure just transition, but this has been a challenge in Taiwan. Although the Air

Pollution Control Act was revised in 2018 where 13 articles in the law stipulated the need for full information disclosure in order to expand citizen participation, TEPA did not however invite citizen participation when drafting the Air Pollution Control Plan, resulting in widespread criticisms during public hearings and among the public at large, when the draft was later released. Taiwan's legislature also proposed amendments to the draft, but the final version adopted by TEPA neither incorporated feedback from public hearings nor proposals by the legislature. Taiwan's air quality governance is therefore still lacking in terms of procedural justice, not only in terms of democratic participation, and the inclusion of science and technology, but also because of the difficulty of holding governance accountable, when it comes to supervising and evaluating the policy impacts.

In terms of *restorative justice*, while the compensation measures in the Air Pollution Control Act were revised, Jobin (2020) and Shih and Tu (2017) pointed to how neoliberal multinational corporations remain the main challenges faced, and the incapacity of legal and social protections to respond to them. Jobin (2020) illustrated this in an example of a collective lawsuit launched against a petrochemical complex in Taiwan:

> *The majority of plaintiffs do not challenge the principle of these patronage payments. But they feel outraged at their uneven distribution, and this accounts for an important motivation in the lawsuit. And feeling justifiably dissatisfied with the company's "good neighbor" practices has made it easier for these residents to start to challenge Formosa Plastics on other, larger issues. Moreover, the victims and their families are not ready to accept that human life counts no more than that of oysters.*
>
> (Jobin, 2020)

It can be seen from the literature that disputes in the adoption and determination of scientific evidence in the courts are therefore one of the most difficult challenges facing Taiwan's restorative judicial practice (Shih and Tu, 2017; Lin, 2019; Liao et al., 2019). In the practice of restorative justice, the inequality of knowledge and capability should therefore require the adoption of the precautionary principle so that "just" compensation is given due consideration and emphasis even before scientific assessments are conducted.

The purpose of *recognition justice* is to allow disadvantaged members of society (usually in terms of wealth or education) to be equally recognized and respected alongside other members of society. Recognition justice, therefore, requires embedding just transition in the social-ecological-technological environments, and thus occupies an important position in promoting air quality governance. While the highly polluted areas in Taiwan urgently require the adoption of differentiated policies, the focus therefore should be placed not just on improving their air quality, but also needs to take into consideration their low incomes and low socioeconomic conditions, and the relative dominance of their agricultural sectors. These counties and cities are therefore being plagued by the rent-seeking behavior perpetuated under neoliberalism, but at the same time, their removal might result in the relocation of industries and increased unemployment and would need strategies to mediate the effects. Similarly, personal transportation comprises a significant component of the expenditure of individuals and families, and more stringent control standards can result in heavier burdens for them, thus while the current scientific assessment is blind to the plight of these groups, justice concerns and considerations would be able to restore their visibility to the governance framework.

Cosmopolitanism refers to the realization of universal values across our societies, and in practice, should transcend nation-state boundaries, ethnicities, class, and wealth. *Cosmopolitanism justice*, therefore, provides the vision for how just transition can be implemented in the face of transnational pollution. A country that allows its high-polluting industries to bring its pollution to other countries will not resolve the global pollution problem, and there is also the question of who to hold accountable in the instance where the pollution produced by one country affects others? Given that clean air should be a universal right, for countries that are forced to accept pollution in their pursuit of food, clothing, and other basic needs, what then should be behind the principle behind whether to tolerate or seek consensus over air pollution risks? This question is one confronting Taiwan which on the one hand, faces pollution from other countries, while its transnational petrochemical industry also contributes to pollution in other countries. However, not only did Taiwan's government not enact green financial policies to rein in these petrochemical industries, it has instead provided extremely cheap water and electricity to help these industries generate profit, while at the same time allowing them to participate in the power generation business, thereby allowing them to become a highly polluting coal-fired power generation company, via cogeneration, or the use of combined heat and power.

The consideration of *space and time* is a necessary parameter of just transition. On the one hand, more in-depth science is needed to support the formulation of differential policies to reduce inequality, while on the other hand, space and time offer clues as to the effectiveness of policy tools. "Space" is directly related to distributional justice, while "time" is directly related to procedural justice. Restorative justice is needed in order for the timely and earnest restoration of each damaged life. Recognition justice is relevant in terms of the space and time for policy implementation, and the time required to do so. The space and time considerations of just transition as such extends to all corners of global geography and history and thus become grounded in the concepts of cosmopolitanism justice and generational justice. The most important concept of all is that no matter the type of justice being practiced, "justice" cannot only matter for specific regions and groups, nor should it only be realized at a specific time. Therefore, when viewed using the principles of just transition, it can be seen that Taiwan's air quality policy has multiple flaws.

7.6 Concluding remarks: the just transition deficit in the environmental governance of post-developmental states

Air pollution is a contemporary consequence of human social, economic and cultural activities, which means that air quality governance is, therefore, an issue embedded in the social-ecological-technological environment (Kasperson and Moser, 2017: 1–2). For this reason, the discussion of the long-term policy consequence of environmental governance in relation to air quality requires an understanding of national development grounded in an awareness of social science. Two issues in the social sciences, "Neoliberal Developmentalism" and "Justice," have also emerged as important issues. The improvement of air quality is therefore a key issue as the post-developmental states undergo their current paradigm shift in the industrial, energy, transportation, and social sectors. If Taiwan can resolve its air pollution problem and achieve the air quality standards set by the World Health Organization, it will certainly lead to major transformations in Taiwan's economic and social developmental model. Since the 1980s, Taiwan has shifted from being a developmental state to neoliberalism; however, it has still not established a governance tradition of environmental and social justice (Hsia, 2015, 2019; Chou, 2017). The main issue of Taiwan's environmental governance is therefore about how to overcome developmentalism in order to move toward sustainability and justice (Chou and Walther, 2016).

This study has shown that the air quality and improvement in Taiwan is unequally distributed while highly polluted areas have seen slower progress. Areas with lower incomes and education, higher crude death rates, and higher agricultural production have also been shown to have higher PM2.5 pollution. Clearly, Taiwan's air quality policy is not aligned to the requirements of a just transition. Therefore, this study proposes that other than developing Taiwan's EPA policies based on air quality models, policies should also be differentiated in accordance with local social and economic conditions in order to achieve a just transition.

The Just Transition framework (Heffron and McCauley, 2018) provides a reference point as to how Taiwan's air quality policy can be evaluated. *Distributional justice* faces obstacles in Taiwan due to the obstruction by neoliberal developmentalism, where the rent-seeking behavior by polluting industries in the low-income counties with higher agricultural production has led to greater air pollution, which has been difficult to overcome due to the inequality in knowledge and power. In terms of *procedural justice*, while Taiwan's air quality policy emphasizes the need to be scientific and evidence-based, policies developed however lacks accurate regional differentiation; and the *recognition of justice* is also seriously lacking among the various stakeholders, which therefore makes it difficult for regulations to be strengthened, such as to curb emissions in the transportation industry. The vision of *cosmopolitanism justice* has also been difficult to achieve due to the extremely cheap water and electricity being provided by the government to help the petrochemical industries profit, and its allowance for them to participate in the power generation business to become a coal-fired power generation company, while at the same time not proactively enacting green financial policies to curb the pollution produced by these petrochemical industries.

This study, therefore, concludes that "just transition" is useful in the study of air quality governance, because it can help to uncover the key issues of developmentalism and injustice in Taiwan's neoliberalism, but more importantly, it forces the government to confront the issue of inequality perpetuated by these models and helps to overcome the impediments of Neoliberal Developmentalism. The just transition framework, therefore, allows the issue of air quality governance in Taiwan to be broken down, so that targeted improvement plans could be developed. The analysis of this research, therefore, points to the introduction of "just transition" as a viable next step in the environmental governance of post-developmental states.

Notes

1 According to Bresser-Pereira (2019), classical developmentalism reflected the conditions and challenges confronted by underdeveloped countries after the Second World War; new developmentalism is a theory based on the successful experiences of growth of middle-income countries. In the mid-2000s, "new developmentalism" started to be formulated under the global trend of neoliberalism. In this article, the new developmentalism imply to the developmentalism under the global trend of neoliberalism.
2 Sustainable environmental governance is governance that ensures ecosystem integrity, economic efficiency, equity, and political legitimacy (Adhikari and Baral, 2018).
3 Due to geographical and climatic differences, the 3 outer island sites are not included.
4 Exclude sites located in the National Park.
5 The outlying islands are not included in this analysis.

References

Adhikari, S., and Baral, H. (2018). Governing Forest Ecosystem Services for Sustainable Environmental Governance: A Review. *Environments*, 5(5), 53.

Ban, C. (2013). Brazil's Liberal Neo-developmentalism: New Paradigm or Edited Orthodoxy? *Review of International Political Economy*, 20(2), 298–331.

Banzhaf, S., Ma, L., and Timmins, C. (2019). Environmental Justice: The Economics of Race, Place, and Pollution. *Journal of Economic Perspectives*, 33(1), 185–208.

Bennett, N., Blythe, J., Cisneros-Montemayor, A., Singh, G., and Sumaila, U. (2019). Just Transformations to Sustainability. *Sustainability*, 11(14), 3881.

Bresser-Pereira, L. C. (2019). From Classical Developmentalism and post-Keynesian Macroeconomics to New Developmentalism. *Brazilian Journal of Political Economy*, 39(2), 187–210.

Chang, Kyung-Sup (2012). Predicaments of Neoliberalism in the Post-developmental Liberal Context. In *Developmental Politics in Transition*. London: Palgrave Macmillan, 70–91.

Chen, Y. L. (2005). Provision for Collective Consumption: Housing Production under Neoliberalism. In R. Y. W. Kwok (Ed.), *Globalizing Taipei: The Political Economy of Spatial Development*. London and New York: Routledge, 99–119.

Cheng, T. J. (2014). Political regimes and development strategies: South Korea and Taiwan. In Manufacturing Miracles (pp. 139-178). Princeton University Press.

Cheng, Y. L., and Li, De-Hsing (2010). Neoliberalization, State and Housing Market Transformation of Public Housing Policies in Taiwan. *Journal of Geographic Science*, 59, 105–131.

Chou, Kuei-tien (2017). *Sociology of Climate Change: High Carbon Society and Its Transformation Challenge*. Taipei: National Taiwan University Press.

Chou, Kuei-tien, Tu, Wen-ling, Walther, David, Shih, Chia-liang, Liou, Yi-ting, and Chao, Chia-wei. (2018). *Taiwan's Air Quality Governance Outlook*. Taipei: RSPRC. http://rsprc.ntu.edu.tw/fordownload/10701/0117/2018working%20Paper.pdf

Chou, Kuei-tien, and Walther, David (2016). *Who is the Turn to Sacrifice? Environmental, Ecological and Social Risks in the Process of Taiwan's Economic Development. Development Studies and Contemporary Taiwan Society*. Chuliu Press: Taipei.

Chou, Kuei-tien, Walther, David, and Liou, Hwa-meei* (2019). The Conundrums of Sustainability: Carbon Emissions and Electricity Consumption in the Electronics and Petrochemical Industries in Taiwan. *Sustainability*, 11(20), 5664–5687 (The authors have equal contributions).

Cohen, A. J., Brauer, M., Burnett, R., Anderson, H. R., Frostad, J., Estep, K., . . ., Forouzanfar, M. H. (2017). Estimates and 25-year trends of the global burden of disease attributable to ambient air pollution: an analysis of data from the Global Burden of Diseases Study 2015. The Lancet, 389(10082), 1907-1918. doi: 10.1016/S0140-6736(17)30505-6

Chu, Y. H. (2007). Re-engineering the Developmental State in an Age of Globalization: Taiwan in Defiance of Neoliberalism. In *Neoliberalism and Institutional Reform in East Asia*. London: Palgrave Macmillan, 91–121.

EEA. (2018). *Unequal Exposure and Unequal Impacts: Social Vulnerability to Air Pollution, Noise and Extreme Temperatures in Europe*. Luxembourg: European Environment Agency.

Féliz, M. (2012). Neo-developmentalism: Beyond Neoliberalism? Capitalist Crisis and Argentina's Development since the 1990s. *Historical Materialism*, 20(2), 105–123.

Franssen, M.-M., and Holemans, D. (2020). Climate, Jobs and Justice for a Green and Socially Just Transition. *GEF Project Just Transition—Working Paper*. Belgium: OIKOS.

Gambhir, A., Green, F., and Pearson, P. J. (2018). "Towards a just and equitable low-carbon energy transition. Grantham Institute Briefing Paper, 26," Retrieved September 9, 2021, from https://www.labor4sustainability.org/articles/climate-jobs-and-justice-a-plan-for-a-just-transition-to-a-climate-safe-economy/

Gao, B. (2013). Neoliberal and Classical Developmentalism: A Comparative Analysis of the Chinese and Japanese Models of Economic Development. In X. Huang (Ed.), *Modern Economic Development in Japan and China*, 68–97. Palgrave Macmillan, London.

Han, S. K., Lee, S. J., and Sung, Y. J. (2014). Shifting Focus in Development Studies: Papers in Development and Society, 1998-2013. *Development and Society*, 43(1), 59–80.

Hajat, A., Hsia, C., & O'Neill, M. S. (2015). Socioeconomic Disparities and Air Pollution Exposure: A Global Review. *Current Environmental Health Reports*, 2(4), 440–450.

Harvey, D. (2005). *The New Imperialism*. Oxford: Oxford University Press.

Harvey, D. (2007). *A Brief History of Neoliberalism*. Oxford: Oxford University Press.

Hashimoto, T. (2014). Discourses on Neoliberalism in Japan. *Eurasia Border Review*, 5(2), 99–119. http://hdl.handle.net/2115/57859

Heffron, R. J., and McCauley, D. (2018). What is the 'Just Transition'? *Geoforum*, 88, 74–77.

Heo, I. (2015). Neoliberal Developmentalism in South Korea: Evidence from the Green Growth Policymaking Process. *Asia Pacific Viewpoint*, 56(3), 351–364.

Heyen, D. A., Menzemer, L., Wolff, F., Beznea, A., and Williams, R. (2020). Just transition in the context of EU environmental policy and the European Green Deal. https://ec.europa.eu/environment/enveco/growth_jobs_social/pdf/studies/just_transition_issue_paper_final_clean.pdf

Hochstetler, K., and Montero, A. P. (2013). The Renewed Developmental States: The National Development Bank and the Brazil Model. *Journal of Development Studies*, 49(11), 1484–1499.

Hsia, Chuan-Wei (2015). *The Neoliberal Turn: The Metamorphosis and Challenge of Developmental State in Taiwan* [Doctoral Dissertation, Institute of Sociology, National Tsinghua University]. Doi:10.6843.NTHU.2015.00529

Hsia, Chuan-Wei (2019). Taiwan's Neoliberal Moment: Crisis, Paradigm Competition and the Ascent of Neoclassical Economists. *Taiwanese Journal of Sociology*, 66, 55–124.

Jobin, P. (2020). Our 'Good Neighbor' Formosa Plastics: Petrochemical Damage (s) and the Meanings of Money. *Environmental Sociology*, 1–14.

Kasperson, R. E., and Moser, S. C. (2017). Introduction: Risk conundrums in a fast and complex world. In R.E. Kasperson (Ed.), *Risk Conundrums* (pp. 1–10). London: Routledge. https://doi.org/10.4324/9781315665894

Kiiza, J. (2012). New Developmentalism in the Old Wineskin of Neoliberalismin Uganda. In *Developmental Politics in Transition*. London: Palgrave Macmillan, 211–232.

Lee, Y. W. (2008). The Japanese Challenge to Neoliberalism: Who and What is 'Normal'in the History of the World Economy? *Review of International Political Economy*, 15(4), 506–534.

Liao, Zhe-Qiang, Chen Yung-Sen, and Chang Chang-Yi. (2019). Social Construction of Nature—A Study on the Environmental Issues of the Sixth Naphtha Cracker in Yunlin. *Environment and World*, 33, 87–111.

Lin, Hung-Yang. (2019). How is the Local Made Not to Work? The Scientific War for and against the No.6 Naphtha Cracker Complex and the Geography of Trust of Mailiao Residents. *Journal of Geographical Science*, 93, 35–80.

Liow, E. D. (2012). The Neoliberal-developmental States: Singapore as Case Study. *Critical Sociology*, 38(2), 241–264.

McNeish, J. A. (2008). Beyond the Permitted Indian? Bolivia and Guatemala in an Era of Neoliberal Developmentalism. *Latin American and Caribbean Ethnic Studies*, 3(1), 33–59.

Mohai, P., Lantz, P. M., Morenoff, J., House, J. S., and Mero, R. P. (2009). Racial and Socioeconomic Disparities in Residential Proximity to Polluting Industrial Facilities: Evidence from the Americans' Changing Lives Study. *American Journal of Public Health*, 99(Suppl 3), S649–S656. https://doi.org/10.2105/AJPH.2007.131383

Morello-Frosch, R., and Jesdale, B. M. (2006). Separate and Unequal: Residential Segregation and Estimated Cancer Risks Associated with Ambient Air Toxics in U.S.

Metropolitan Areas. *Environmental Health Perspectives*, 114(3), 386–393. https://doi.org/10.1289/ehp.8500.

Polanyi, K. (1944). *The Self-regulating Market and the Fictitious Commodities: Labor, Land and Money.*

Radice, H. (2008). The Developmental States under Global Neoliberalism. *Third World Quarterly*, 29(6), 1153–1174.

Ramesh, M. (2003) Globalization and Social Security Expansion in East Asia. In L. Weiss (Ed.), *States in the Global Economy: Bring Domestic Institutions Back In.* Cambridge and New York: Cambridge University Press, 83–100.

Shih, Chia-Liang, and Tu, Wen-Ling. (2017). The Scientific Framework and Decision Deadlock in the Environmental Administrative Procedures: Examining the EIA of the Fire Accident in the No. 6 Naphtha Cracking Project, 2010. *Journal of Public Administration*, 52, 81–111.

Suzuki, M., Ito, M., Ishida, M., Nihei, N., and Maruyama, M. (2010). Individualizing Japan: Searching for Its Origin in First Modernity. *The British Journal of Sociology*, 61(3), 513–538.

TEPA. (2021). Air quality Annual. TEPA (Environment protection agency, Taiwan). April, 2021, Retrieved from https://www.epa.gov.tw/DisplayFile.aspx?FileID=6BB17C4B4AD72C76

Tessum, C. W., Paolella, D. A., Chambliss, S. E., Apte, J. S., Hill, J. D., & Marshall, J. D. (2021). PM2. 5 polluters disproportionately and systemically affect people of color in the United States. Science Advances, 7(18), eabf4491.

Tu, Wen-Ling, Chou, Kuei-tien, and Walther, David. (2017). *Research on the Risk Governance and Regulation of Taiwan Air Pollution.* Taipei: RSPRC.

Tu, Wen-Ling, and Shih, Chia-Liang. (2019). Challenging Air Pollution: An Explorative Study of Community Action Science and Localized Practices. *Journal of Communication Research and Practice*, 9(1), 1–32.

UN. (2019). Issue of human rights obligations relating to the enjoyment of a safe, clean, healthy and sustainable environment. Human Rights Council, UN. April, 2021, Retrieved from https://digitallibrary.un.org/record/3814570

UNEP. (2019). Air Pollution in Asia and the Pacific: Science-based Solutions. UNEP. https://wedocs.unep.org/handle/20.500.11822/32101

Walther, David, and Chou, Kuei-tien. (2018). Insufficient Trans-boundary Risk Knowledge and Governance Stalemate in Taiwan's PM2.5 Policy-making. *10th Annual Conference on Development Studies: Taiwan Experience 2.0: Global and Local Development*, October 28, NCCU.

Weiss, L., & Hobson, J. M. (1995). States and economic development: a comparative historical analysis. Polity Press.

Wong, J. (2006). *Healthy Democracies: Welfare Politics in Taiwan and South Korea.* New York: Cornell University Press.

8 Air quality governance in China

Authoritarian environmentalism and the coal-to-gas switch campaign

Chelsea C. Chou, Chih-sung Teng, Chin-en Wu and Kuan-chen Lee

8.1 Introduction

On October 18, 2017, at the opening of the Nineteenth Chinese Communist Party (CCP) Congress, Xi Jinping emphasized that "building an ecological civilization" was critical for the country's sustainable development. To put into practice the top leader's doctrine, varieties of government policies have been prepared to win the battle to deliver blue skies. Among them, one of the most important campaigns was to reduce the use of coal. In order to speed up the process, on December 5, 2017, the Chinese government promoted the "Clean Winter Heating Plan in Northern China (2017–2021) (关于印发北方地区冬季清洁取暖规划的通知)" (National Development and Reform Commission et al., 2017) and launched the "Coal-to-Gas Switch" campaign in its northern provinces, including "2 + 26 cities," namely two direct-administered municipalities, Beijing and Tianjin, and twenty-six prefecture-level cities across Hebei, Shanxi, Shandong, and Henan province.[1] This campaign is sometimes abbreviated by the media as the "Beijing-Tianjin-Hebei (BTH) Goal-to-Gas Switch (京津冀煤改气)" as the twenty-eight cities are located around the capital area.

The major step of the campaign was to stop all small coal-fired boilers and replace them with natural gas. To achieve the goal, the Chinese government, along with its environmental experts, formulated the discourse that everyone was creating air pollution and persuaded ordinary local residents and small business owners to dump their coal stoves and to stop using coal-generated heat. This resulted in a freezing winter for many households in December 2017 and January and February 2018. In this chapter, we examine the formulation of this "Coal-to-Gas Switch" campaign and explore the process of how the Chinese government intentionally individualized the pollution responsibility in order to deflect public criticisms from large state-owned enterprises (SOEs) and the government. We argue that while the authoritarian Chinese government has made up its mind to eliminate air pollution, the non-participation of social actors in its policymaking has in fact led to biased polices which are eventually unsustainable. These policies merely focus on individuals' responsibility in

DOI: 10.4324/9781003211747-11

protecting the air quality and neglect the problems caused by the state-owned heavy industry. In other words, the authoritarian government might be decisive in responding to air pollution, but the biased polices have made it difficult to improve the air quality in the long run. As we will show, the government had to slow down the campaign to switch from coal to gas when insufficient heating was widely reported in the northern cities. This caused the air pollution degree to rise again.

To evaluate the effectiveness of this "Coal-to-Gas Switch" campaign, we collect all the data from China's local monitoring stations from 2015 to 2109 to check the ups and downs of China's particulate matter 2.5 (PM2.5) level. In 2017, there were roughly 1,600 stations across the country. We calculate each prefectural city's level of air quality and compare the reduction of air pollution between the Beijing-Tianjin-Hebei region and the rest of China throughout the five years from 2015 to 2019. We find that the campaign was indeed successful in reducing PM2.5 in the beginning. Our data shows that in the twenty-eight cities in the BTH region, the degree of PM2.5 was significantly decreased in late 2017 and early 2018. This decrease was particularly salient when comparing the degree of PM2.5 in late 2016 and early 2017. However, the degree of PM2.5 started to rise again in late 2018 and early 2019. This shows that while the campaign was effective when first introduced in December 2017, due to the lack of understanding of the local gas facilities, the policy was not sustainable. Moreover, the resurgence of air pollution also shows that while the Chinese government had for many years made efforts to improve the clean energy of state-owned heavy industries, the problem had not been completely resolved. Without fully addressing the pollution problem caused by the household use of coal and the state-owned heavy industry, China's air quality could hardly have foundational improvements.

The chapter proceeds as follows. The next section discusses the theoretical aspect of China's particular type of policymaking in combating environmental problems. We argue that the authoritarian government did not succeed in bringing blue sky due to the biased policy regulation. The third section details the 2017 BTH Coal-to-Gas Switch campaign and finds that the intent to individualize the pollution responsibility led to governance failure as the campaign was not sustainable. In the fourth section, we provide a visual presentation of China's changing air quality in different localities. We demonstrate that the air quality started to deteriorate again when the Coal-to-Gas Switch campaign was forced to stop. The chapter concludes in the fifth section.

8.2   Authoritarian environmentalism in China

China's approach to air quality governance has long been described as authoritarian environmentalism (Gilley, 2012; Li et al., 2019). Are authoritarian governments more effective than democracies in solving environmental problems? Advocates of authoritarian environmentalism argue that facing severe environmental challenges, authoritarian governments are autonomous to exclude enterprises and business actors when formatting environmental policies (Wells, 2007). This exclusion is important to avoid interference from the business sector so that the government can focus on combating pollution without worrying about hurting the interest of business community. In addition to making environment-friendly policies, authoritarian governments are able to achieve better efficiency in implementing these policies. By restricting individual liberty, authoritarian governments have the capacity to regulate societal actors' unsustainable behavior that would undermine the implementation of environmental protection rules (Beeson, 2010). In one word, either in the policymaking or implementation process, authoritarian governments are better equipped with tools to overcome environmental challenges.

While many people believe in the effectiveness of authoritarian environmentalism, empirical research on China has long found that environmental problems have not been adequately dealt with under authoritarian rule. Most studies focus on the adverse circumstances during the policy implementation process and point out that the decentralized and fragmented central-local relations in China have significantly undermined the practice of authoritarian environmentalism (Gilliey, 2012). Although China is an authoritarian regime, fragmentation in the state apparatus is never a new phenomenon (Lieberthal & Oksenberg, 1988; Lieberthal & Lampton, 1992). Since the reform and open-door policy was introduced in the 1980s, in order to respond to increasingly complicated economic affairs, more decision-making power has been delegated to professional ministries within the State Council and to local governments at different levels. This has created space for bureaucratic bargaining and negotiations among ministries and between central and local governments in the Chinese governing system. In the policy implementation process, these loopholes have given considerable discretion to local governments.

In other words, the central government relies on local governments to carry out its environmental policies, but local officials' compliance is always a problem. To respond to this, the Chinese central government under Xi Jinping has made enormous efforts to re-centralize its environmental governance structure (Kostka & Nahm, 2017). For example, the cadre promotion criteria are reconstructed, and the cadre turnover system is installed to evaluate local officials more frequently. Reducing local autonomy is found to be associated with the success of "green urbanization" (Chen & Lees, 2018), but high cadre turnover has incentivized local officials to adopt quick but cheap and even unsustainable steps to demonstrate their performance in protecting the environment (Eaton & Kostka, 2014). In practice, local officials with short time horizons are likely to hinder the full implementation of central government's agenda of environmental protection. Accordingly, while being an authoritarian regime, fragmentation in the policy enforcement process has given reason to doubt the effectiveness of China's authoritarian environmentalism. The policy implementation does not gain the advantage from being an authoritarian regime.

Accordingly, improving the quality of China's environmental governance requires local governments' engagement and collaboration. However, as will be shown in the next section, local policy implementation is not the only obstacle to China's environmental protection. While aspiring to be eco-friendly, the authoritarian central government does not always make sound and sustainable environmental policies. Some of the central policies are in fact detrimental to environmental protection in the long run. Therefore, while many local officials are likely to delay the full implementation of central environmental agenda, the central government itself is also the answer to why environmental problems have not been adequately dealt with under the Chinese authoritarian governance.

As mentioned, excluding the business sector in the decision-making process enables authoritarian governments to announce policies focusing on environmental problems. But this does not mean that authoritarian governments do not need to take into consideration the interest of business sector when making their environmental policies. Every government must manage multiple and sometimes even competing objectives, and the authoritarian government is no exception. After all, the major concern of authoritarian governments is their regime survival and stability. While many quality-of-life issues including environmental protection have become one of the major concerns for the Chinese people (Wu et al., 2017), maintaining a high degree of economic growth is still the critical way for the Chinese government to gain political legitimacy (Zhu, 2011). As a result, when drafting environmental policies, the Chinese government needs to consider any possible short-term economic shocks. This consideration is particularly important today as many large heavy industries that are prone to cause pollution are state-owned enterprises (SOEs). In the Xi Jinping era, SOEs' importance in leading the county's economic development has been frequently emphasized (Livingston, 2020). For the Chinese government, the best scenario is that environmental problems can be addressed without harming the development of state-owned heavy industries.

To downplay the role of SOEs in generating pollution, the Chinese government deliberately articulates discourses that indicate the responsibility of every citizen for protecting the environment (保护环境人人有责). This helps the authoritarian government to keep the developmental state priorities of preserving state-owned energy-intensive industries. Scholars of Chinese policymaking have already pointed out that the Chinese government has regularly actively engaged and even mobilized citizens when drafting its policy initiatives (Huang & Yang, 2002). This engagement does not only involve creating propaganda schemes or slogans to foster public compliance but also includes actions to persuade ordinary citizens to bear the responsibility in carrying out government policies. In environmental policymaking, when pollution deteriorates and public criticisms occur, it is in the Chinese government's interest to focus on social actors' responsibility because it deflects people's attention from criticizing the government and the business sector. Such a strategy brings ordinary people into the campaign and prevents the government and the state-owned heavy industry from being the target of blame.

In the next section, we use China's combat of air pollution as an example to illustrate how the Chinese government formulates policies to mobilize ordinary individuals to share the responsibility without giving away its developmentalist agenda. We find that since 2017, the Chinese government has gradually initiated the coal-to-gas switch campaign to reduce the consumption of scattered coal. This policy was endorsed by ten central ministries including the National Development and Reform Commission, the then Ministry of Environmental Protection (now the Ministry of Ecology and Environment), the National Energy Administration, etc. The "Clean Winter Heating Plan in Northern China (2017–2021)" was issued on December 5, 2017, to facilitate the task (National Development and Reform Commission et al., 2017). According to this plan, scattered coal which was widely used in households cooking and heating, and small-scale industrial boilers and furnaces was the major source of air pollutant emissions. The critical step to clear air was thus to replace the old boiling and heating system with various new clean facilities in both urban and rural areas. By focusing on the negative impact of civil use of the scattered coal, this policy turned ordinary households and small business industries into the campaign.

When formulating the campaign, the Chinese government articulated discourses with the help of environmental experts to build the policy's credibility. However, as the next section will show, this policy had serious flaws and in fact unnecessarily worsened the living condition of ordinary people. Reducing the use of scattered coal in the heating system led to freezing winter when the new heating facilities had not yet been established. This reveals the inadequacy of policymaking in authoritarian regimes. Lacking media freedom and regular elections, authoritarian governments do not have adequate mechanisms to know the real situation at the grassroots level (Wintrobe, 1998, 2001). The weaknesses in information collection often lead to the failure of prescribing appropriate policies to solve the environmental problem. In other words, in the Chinese authoritarian regime, environmentalism is undermined by both the policymaking and implementation process. While local cooperation is a problem, what the central government has announced is not always sound and proper in combatting environmental pollution.

8.3   The coal-to-gas switch campaign in northern provinces

In the action against air pollution, reducing PM2.5 is one of the most important steps. On September 10, 2013, the State Council issued the "Air Pollution Control Action Plan" (大气污染防治行动计划) to require a nationwide reduction of PM2.5 concentrations by 2017 (State Council, 2013). The specific indicator in this plan included that in the Beijing-Tianjin-Hebei region, Yangtze River Delta, and Pearl River Delta, the level of current fine particles should be reduced by 25%, 20%, and 15%, respectively, and the number of good air quality days should be increased every year. For other cities, the plan asked to reduce more than 10% of PM2.5 concentrations. In addition, the plan also set a standard that the annual average concentration of PM2.5 in Beijing should be controlled at about 60 micrograms per cubic meter (60 μg/m3) by 2017. From the Action Plan, we see that 2017 was a critical year for the Chinese government to succeed in reducing air pollution.

8.3.1 Central and local regulations: small enterprises in focus

To fulfill the goal set in the Action Plan, various government documents were issued. On June 20, 2016, the then Ministry of Environmental Protection announced the "Strengthening Measures for the Prevention and Control of Air Pollution in the Beijing-Tianjin-Hebei Region (2016–2017)" (京津冀大气污染防治强化措施) (Ministry of Environmental Protection et al., 2016). In this document, the timeline was set to implement several required actions. First, before the end of 2016, all the thermal power, steel, and cement enterprises in the Beijing-Tianjin-Hebei region needed to have a permit for emissions of pollutants. Second, all the coal-generated heating systems and old coal stoves should be replaced by the end of October 2017. At the same time, small-scale and scatter-distributed polluting factories should be eliminated. More importantly, the document delineated areas where the use of coal was totally prohibited. It asked to implement the "No Coal Zone" in eighteen county-level administrative districts in Baoding City and Langfang City (Hebei Province) that bordered Beijing Municipality. Within the "No Coal Zone," if the use of coal was still needed, the government should strictly assure its quality while making efforts to reduce the amount as much as possible.

In addition to the regulation issued by the Ministry of Environmental Protection, the State Council included the rules on how to reduce the use of coal in its "Notice of the 13th Five-Year Ecological Environment Protection Plan" (十三五生态环境保护规划通知) (issued on December 5, 2016). The Notice asked to regulate the civil use of scattered coal, to supervise the pollution generated by motor vehicles, to increase the supply of non-fossil energy, and to switch coal to gas and/or electricity (State Council, 2016).

The "No Coal Zone" regulation was reinforced by another round of policy-making. On Feb 17, 2017, the Ministry of Environmental Protection together with other ministries issued the "Work Plan for the Prevention and Control of Air Pollution in the Beijing-Tianjin-Hebei Region and Its Surrounding Areas in 2017" (京津冀及周边地区2017年大气污染防治工作方案) (Ministry of Environmental Protection et al., 2017a). In this document, small enterprises were the focus. The Plan asked local governments to shut down all "small-scale and scatter-distributed polluting enterprises" (小散乱污企业) by the end of October 2017.[2] The local governments were encouraged to use measures such as dismantling production facilities and cutting water and electricity of these enterprises in order to speed up the shutdown. In addition, the grid management system should be installed and the ground-level leaders including those in the townships (乡), towns (镇), and streets (街道) should serve as the "grid leader" to carefully supervise these enterprises. In addition to shutting down these small enterprises, the Work Plan urged the implementation of a clean winter heating system in the 2 + 26 cities in the northern provinces. It stated that in cities where the use of coal had not been totally eliminated, before the end of October, they should replace coal with gas or electricity in more than 50,000 households, and increase the utilization of low-grade industrial waste heat and geothermal energy.

In addition to central regulations, local governments also took measures to underpin air pollution control. The Beijing municipal government, for example, increased subsidies up to 50% to private companies and households when they conducted clean energy transformation (Beijing Municipal Commission of Development and Reform, 2016). During the annual meeting of the National People's Congress, the governor of Hebei province, Qingwei Zhang (张庆伟), mentioned that while the heavy industry was important for the province's economic growth, Hebei had already shut down several major cement plants and would stop building any new power plants. Hebei province had invested 24 billion RMB in air pollution control. It was expected that with these measures, the fine particle pollution in Hebei will improve significantly within three to five years (Wen, 2016).

Despite these efforts, the Beijing-Tianjin-Hebei region was again in heavy air pollution in the winter of 2016. On October 14, 2016, the PM2.5 index in Beijing municipality even reached a peak at 368. It was at the hazardous level while an average of 12.0 µg/m3 or less was generally considered to be satisfactory. On the same day, the PM2.5 index in Tangshan city and Chengde city (Hebei province) reached 297 and 298, respectively, a very unhealthy level (Ye, 2016). In the same month, the Beijing-Tianjin-Hebei region was hit by serious smog again, and the number of smog days in 2016 outnumbered the past decade (Ai, 2016). This suggests that there was still a lot of work to be done to improve the air quality in these northern provinces.

To take remedies for the problem, on April 5, 2017, the Ministry of Environmental Protection launched what it said was the largest inspection in history. More than 5,600 environmental law enforcement officers participated in this inspection in all the "2 + 26" cities. From the beginning of April to mid-May in 2017, the inspection team investigated 11,674 enterprises and found that more than two-thirds of them did not fully comply with the aforementioned environmental regulations. The common violation included abnormal operation of pollution control facilities, fabrication of the monitoring data, and improper industrial dust controls. In addition, the inspection also revealed that small-scale and scatter-distributed polluting enterprises were under-regulated. Many local authorities did not report the actual number of these small enterprises, which significantly increased the central government's difficulty in knowing the real situation (Gao, 2017).

To improve grassroots supervision, on August 21, 2017, the Ministry of Environmental Protection issued a new document, entitled the "2017–2018 Comprehensive Air Pollution Governance and Action Plan for the Autumn and Winter in the Beijing-Tianjin-Hebei Region and Its Surrounding Areas" (京津冀及周边地区2017–2018年秋冬季大气污染综合治理攻坚行动方案) (Ministry of Environmental Protection et al., 2017b). This document indicated that while enterprises should take actions to reduce pollution, local governments were the main supervisor responsible for policy implementation and should be accountable for air quality improvements. More specifically, the Plan strengthened local governments' responsibility in managing the air quality through several concrete steps. First, city and county officials were required to report the progress of pollution control every month; Second, if the improvement of air quality did not meet the target set by the government, an accountability mechanism should be installed to impose punishments to the local official. With these efforts, according to the data released by the Ministry of Environmental Protection, the proportion of enterprises that violated pollution control regulations dropped enormously after the 20th round of inspection in January 2018 (Dong, 2018).

8.3.2 *Bringing everyone into the campaign*

As mentioned, many of the government regulations as well as the central ministry's inspection focused on the air pollution generated by small-scale and scatter-distributed polluting enterprises. According to the Chinese government, big companies were generally less polluted than small enterprises. The Chinese government claimed that since the early 2010s, it had asked big companies to gradually upgrade their pollution control equipment. In 2017, most power plants and big enterprises had already installed the most cutting-edge facilities to reduce pollutions. The remaining problem now was the small coal burners that were mostly used by small factories and households for heat and power. According to the Chinese government, as the level of PM2.5 degrees in winter was much more serious than in summer,[3] traditional heating must be one of the major sources of pollution.

Against this backdrop, the Chinese government emphasized the need to dump small coal burners and included small enterprises and ordinary households in the battle to win blue skies. In March 2016, the Ministry of Environmental Protection issued the "Technical Guidelines for Comprehensively Governing the Use of Scattered Coal and Burning Pollution in Rural Areas" (农村散煤燃烧污染综合治理技术指南) (Ministry of Environmental Protection, 2016a). Then in October 2016, the "Technical Guidelines for Comprehensively Governing the Civil Use of Coal Burning Pollution" (民用煤燃烧污染综合治理技术指南) was announced to specify the details of dumping coal stoves and switching from coal to natural gas and/or electricity for heat and power (Ministry of Environmental Protection, 2016b). These documents marked the beginning of the national campaign aiming at curbing air pollution by incorporating not only the heavy industry but also ordinary people into the fight. The goal of the campaign was to replace coal with electricity or natural gas for 3 million households. In an article published by China Electric Power News, an official newspaper supervised by the Energy Bureau of the National Development and Reform Commission, experts argued that addressing the issue of coal burning for heat was the most urgent problem for air quality governance. It was because small coal-burning boilers, including household stoves and small factory boilers, were much less efficient than most industrial boilers (Ji, 2017). In another article written by China Environment News, a ministry newspaper supervised by the Ministry of Environmental Protection, the government indicated that while small coal boilers accounted for about 33% of the total coal consumption in China, they contributed to more than 60% of the air pollution (Li & Feng, 2017).

These newspapers of the central ministries emphasized the low efficiency of burning coal for heat and urged local governments to push households, especially in the rural area, to switch from coal to gas or electricity. In Shanxi province, a traditional industrial region facing heavy air pollution, its provincial "Air Pollution Action Plan for the Autumn and Winter" centered on eliminating small-scale coal-burning boilers. Among the top list, switching coal to gas for heat and phasing out small and sporadic but highly polluting enterprises were two of the most important steps to curb air pollution (Cheng, 2017).

While both the central and local governments had paid much attention to reducing coal consumption, the results were not satisfactory. The geographical location of small coal boilers was spread across a large area, making it very difficult for the government to regulate them. In order to speed up the process, the Chinese government at all levels illustrated the concept that combating air pollution was a shared responsibility and everyone had to make some efforts. For example, an article published by a local official newspaper writes:

> Most of the sources of air pollution is closely related to our daily life and activities. In fact, each of our inadvertent move may be a "killer" that pollutes the blue sky and becomes a producer of the air pollution. Having a blue sky and clear water is everyone's wish. As we all are the possible source of pollution, protecting the blue sky is a common responsibility of the whole society, and is also a duty of everyone.
>
> (Xu, 2016)

When everyone was a responsible player in the air pollution, it justified the stopping of civil use of coal burning. The government implemented plans to remove all coal combustion in the urban area, and gradually replaced coal with natural gas in mountainous areas. As a result, many old coal stoves were dumped.

With ordinary households' participation, the action to reduce the coal consumption made some achievements. When the winter was approaching in late 2017, knowing that coal was the major source of heat energy in China, the government at all levels found that the process of discarding coal stoves must be accelerated. Otherwise, the air pollution in winter would rapidly escalate. Accordingly, the campaign to dump coal stoves was speeded up in order to end the dependence on coal for heating. However, many coal stoves were removed before new cleaner-burning natural gas furnaces were installed. As a consequence, many local residents in the northern provinces encountered an extremely frozen winter, shivering without enough heat.

8.3.3   *Individual environmental responsibility and developmentalism*

The goal-to-gas switch campaign represents a typical example of how the Chinese government defects public criticism through individualizing the pollution responsibility. According to the 2016 report of Energy Foundation China,[4] in the Beijing-Tianjin-Hebei region, the main source of air pollution included coal burning (34%), motor vehicles (16%), and industrial waste emission (15%). In Beijing, the main PM2.5 emission sources were coal and motor vehicles. In Hebei and Tianjin, coal burning, the chemical industry, and heavy metal smelting were all the main sources of heavy metal pollution (Energy Foundation China, 2016). In addition to policy reports, a bulk of academic research also echoes the same conclusion. According to Tang et al. (2020), heavy industry, vehicle emissions, the burn of loose coal, and dust were all attributors of the severe air pollution in the BTH region. As Holdaway et al. (2018) point out, in Hebei and Tianjin, although most SOEs had gradually improved their filter system, emissions from the heavy industry were still substantial. In the estimation of Bo et al. (2017), the percentage of pollution produced by the steel plants in the BTH region was as high as 14.0% in winter and 13.1% in summer.

Although the heavy industry was an important factor of pollution in the BTH region and households and small businesses were just one of the sources of total coal consumption, the Chinese government emphasized on many occasions that these users did not have cleaning and filtering systems, as did large enterprises and power plants. The emission technology of coal-fired power plants and large enterprises to discharge pollutants was more advanced than the smaller players. Accordingly, eliminating coal use by households and small businesses would contribute more to the reduction of coal emissions. This served as the justification for the campaign of dumping coal stoves. By articulating the discourse that civilian use of coal was dangerous to the environment, the Chinese government mobilized the mass to share the responsibility.

Focusing on the problem caused by households and small enterprises helped the Chinese government to leave the fundamental issue untouched. The dilemma of China's air quality governance was that China's economic growth strongly relied on government spending and corporate investments in various infrastructure projects (Sutter and Sutherland, 2021). This investment-driven growth model typically involved the development of heavy industries including the steel, cement, and chemical sector. In the past decades, the Chinese government had made efforts to upgrade the industrial structure and to restructure its economy but the outcome was not very successful. At the local level, although environmental protection had been included on the list of cadre promotion, many local governments still relied on traditional infrastructure projects to boost revenue and local economic growth. Local governments thus still kept a blind eye on the overcapacity of these projects. Even if many of the big factories had upgraded their waste treatment system, the total amount of pollution was still difficult to cap (Modern Express, 2013).

At the same time, the central government was in fact reluctant to abandon the heavy and chemical industries as preserving state-owned energy industries was one of its developmentalist priorities. In the Xi Jinping era, there was a resurgence of emphasis on SOEs. The central role of SOEs in the economy was unabashedly endorsed. In his speech for the Nineteenth CCP Congress in October 2017, Xi pointed out that it was necessary to strengthen the role of SOEs, to optimize its structure, and to promote SOEs to be stronger and bigger. For environmental protection, Xi mentioned the need to enforce stricter pollutants discharge standards. When outlining the plan to solve prominent environmental problems, Xi himself also emphasized the importance of getting everyone involved in the battle against pollution. From the Chinese government's perspective, in a situation where industrial upgrading remained sluggish and the dominance of SOEs and heavy industries was unlikely to change anytime soon, promoting the concept of individual environmental responsibility was particularly important to improve the air quality governance. Accordingly, while upgrading the pollution prevention equipment of power plants and big companies were still on the agenda, curbing small household stoves and erasing sporadic small polluters became the centerpiece of government pollution prevention policies.

Focusing on the household use of coal was helpful in deflecting people's attention from criticizing state-owned heavy industries for producing pollutions. Compared to the light industry and electronics manufacturing, heavy industries were mostly energy-intensive, relying more on coal power plants and hence creating higher levels of carbon emissions. As coal was still the dominant source of power, a high concentration of heavy industries tended to go along with higher levels of fine particles in the air, even if the power plants had filtering equipment. Heavy industries had long been important for China's economic growth. The most recent craze for the development of heavy industry occurred in the early years of the twenty-first century. Take Tangshan City in Hebei Province as an example. According to the official data published by the City's Statistical Bureau, in 2016, the five leading industries included the steel, energy, chemical, construction material, and equipment manufacturing industry (Tangshan Statistical Bureau, 2017). Although there were overcapacity problems in the steel industry, it was hard for the Chinese government to make the decision to restrict the production of steel. As heavy industry was one of the main pillars of China's economy, for developmentalist concerns, it was not in the government's interest to let the heavy industry become the target of blame. As reducing the heavy industry was not a good option, the government turned to the second solution.

However, this policy could not be implemented for a long time. Before 2017, most government newspapers unilaterally praised the benefits of abandoning coal stoves, but since the policy caused insufficient heating, many newspapers in 2018 started to create slogans such as "systemic thinking needs to be emphasized when making decisions," "combatting the pollution should not let children get cold," and "one-size-fits-all approach is not suitable for coal reform," etc. These narratives indicate that the Chinese government had realized that this policy must be slowed down. As will be shown in the next section, the air quality began to deteriorate when the policy of abandoning small coal stoves was suspended. Such results reveal that the policy was not sustainable and the Chinese government needed to speed up the installation of natural gas infrastructure to transport gas to homes before putting into practice the coal-to-gas switching policy. Moreover, the rising air pollution also shows that if the Chinese government continue to only focus on the responsibility of households and small enterprises and do not comprehensively review all the source of air pollution, it is difficult to succeed in air quality governance.

8.4 Was the coal-to-gas switch campaign effective in reducing air pollution?

With the discourse to emphasize the responsibility of households and commercial users, the campaign for dumping small coal boilers was widely implemented in northern China. After all, was it helpful in improving the air quality? Several international NGOs reported that the Coal-to-Gas Switch Campaign was effective in combatting air pollution in late 2017. According to the report published by Greenpeace, the concentration of PM2.5 dropped 33.1% in Beijing, Tianjin, and the 26 surrounding cities in the last three months of 2017 (Greenpeace East Asia, 2018). In this section, we examine the ups and downs of China's PM2.5 levels to see if utilizing less coal-generated heat during the winter helped reduce air pollution.

To have a grasp of China's changing air quality across different regions, we collect all the data from China's ground monitoring stations.[5] As in 2017, there were 1,563 stations across the country (Figure 8.1).[6] These stations were directly supervised by the Ministry of Environmental Protection.

Based on the data of these local monitoring stations, we calculate each prefectural city's PM2.5 value. We conduct inverse distance weighting (IDW)

Figure 8.1 The geographical distribution of ground monitoring stations in China, 2017.

interpolation on the observations of all the ground monitoring stations to estimate the unknown value of each city from the nearby known points. The original PM2.5 observations of these local stations were reported hourly on a daily basis. We calculate the average of these numbers for the 24 hours of a day to obtain each city's daily value. Below we provide a visual comparison of the PM2.5 level in summer and winter in 2017.

From Figure 8.2 we see that the degree of air pollution was indeed much more serious in winter than in summer. This is one of the major reasons why

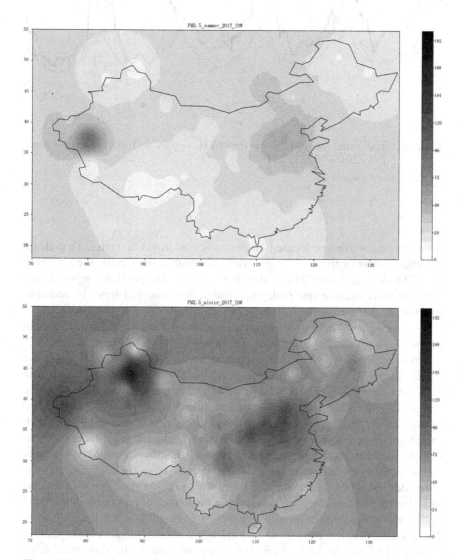

Figure 8.2 The concentration of PM2.5 in summer versus winter in 2017

Source: Authors' calculation

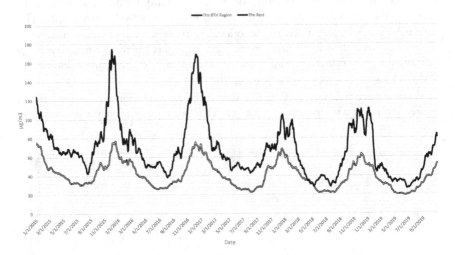

Figure 8.3 The value of PM2.5 in the BTH region and the rest of China, 1/1/2015–10/31/2019

Source: Authors' calculation

the Chinese government focused on dumping coal stoves in winter. To evaluate the effectiveness of the coal-to-gas switch campaign, we examine the evolution of PM2.5 degrees in the "2+26 cities" in the Beijing-Tianjin-Hebei region across different years because this policy was mainly implemented here. In addition, we compare the air quality in the BTH region with the rest of China to see if the former achieved better performance in reducing PM2.5 values with the help of the coal-to-gas switch campaign. The result is shown in Figure 8.3.

From Figure 8.3 we see that compared to late 2016 and early 2017, the PM2.5 in the BTH region (the black solid line) decreased dramatically in late 2017 and early 2018. The PM2.5 value in December 2016 was almost at the level of 170 µg/m3, but had been significantly reduced to the level of 80 µg/m³ in December 2017. This huge difference shows that the coal-to-gas switch campaign along with the large-scale inspection in the northern cities was successful in combatting the air pollution. Individualizing the pollution responsibility was so persuasive as to mobilize ordinary people to join the campaign.

In addition, when comparing the air quality in the BTH region with the rest of China (the hollow line), we also see that the degree of reduction in the BTH region was much higher than in the rest of China. While the hollow line in January 2018 (at the level of 55 µg/m3 approximately) is also lower than in January 2017 (at the level of 65 µg/m³ approximately), the downward slope of the solid line is much steeper. This comparison provides another piece of evidence indicating that with the help of the coal-to-gas switch campaign, the

BTH region could achieve better performance in reducing PM2.5 concentrations.

However, we see from Figure 8.3 that in late 2018, the PM2.5 in the BTH region started to rise again. Compared to November and December 2017 when the daily concentration of PM2.5 was about 85 μg/m³ in the BTH region, in November and December 2018 the daily PM2.5 value increased to 95 μg/m³ and 100 μg/m³. In addition, in January and February 2019, the PM2.5 value in the region was in general bigger than in January and February 2018. The number of days (about 16 days) when the PM2.5 value was higher than 95 μg/m3 were also more in early 2019 than one year earlier in early 2018 (about 6 days). This shows that at the beginning of the coal-to-gas switch campaign, China's air pollution was substantially eased, but after one year, we do not see the improvement being kept.

While the level of PM2.5 concentrations in late 2018 did not worsen back to the level in late 2016, the Chinese government in 2018 was not able to maintain good air quality in late 2017. Reducing coal-generated heating helped deliver better air quality in the beginning, but when this campaign caused insufficient heating for ordinary households, the government had to slow down the campaign. As a result, the level of PM2.5 concentrations increased in the BTH region and it was impossible to maintain the good air quality as last year.

Accordingly, while reducing the household use of coal was one of the necessary steps to combat air pollution, the goal could not be achieved before the installation of natural gas infrastructure. But the problem was that the central government had not been fully aware of the shortage of local gas supply. Moreover, as pointed out by the previous section, the Chinese government did not pay as much attention to state-owned heavy industries. Although heavy industry was also one of the major sources of pollution, switching the household use of coal was the focus in the BTH region. This shows that the Chinese policymaking was biased toward SOEs.

In a nutshell, while bringing everyone into the campaign and focusing on the responsibility of small enterprises helped speed up the process of dumping small coal burners, the policy was unsustainable due to the government's lack of understanding of the situation about gas supply at the grassroots level. In other words, the coal-to-switch campaign was initiated at a time when the infrastructure of gas heating had not been sufficiently installed. The central government failed to know the process. In fact, many local households still heavily depended on coal for heating during the winter. The success of PM2.5 reductions made by the no-coal campaign had brought about problems in other areas of people's livelihood. But only when insufficient heat was causing suffering to its people during the freezing winter did the Chinese government see the unsustainability of switching to gas. These problems forced the government to ease the campaign which eventually drove the degree of PM2.5 to increase again.

Therefore, individualizing the pollution responsibility was a deliberate strategy employed by the authoritarian regime to combat air pollution when economic development was the priority. To enforce individuals' cooperation, the Chinese government broadcasted the propaganda of everyone's responsibility and the strategy was successful in the beginning. Unlike in a democratic setting where unpopular policies would quickly lead to societal criticisms and civil disobedience, the decision-making of authoritarian environmentalism could be very effective. However, individualizing the pollution responsibility was not a sustainable policy tool. The problem was that without adequate channels to know the insufficiency of gas supply at the grassroots level, the authoritarian government did not make the right decision to implement the switching to gas policy. Defects of the authoritarian policymaking system caused unsustainable and biased air governance.

8.5   Conclusion

This chapter examines China's air quality governance through an analysis of the 2017 "Coal-to-Gas Switch" campaign. To accomplish the goal of switching to natural gas, the Chinese government formulated discourses to encourage individuals to abandon small coal-fired boilers. Slogans focusing on individuals' responsibility for air pollution were created to facilitate the task. In addition, as scattered coal was difficult to monitor due to the fact that it was widely used across the northern area, the Chinese government highlighted that the coal stoves used by households and small businesses for heating in winter were the major source of air pollution. With the help of these discourses, the Chinese government was able to defect itself from being the target of accusations. Moreover, emphasis on the pollution produced by ordinary households and small enterprises helped circumvent the responsibilities of large state-owned enterprises. For the Chinese government, when maintaining the developmentalist goal was at the top of its governance agenda, focusing on individuals' responsibility in combatting air pollution was the best strategy without harming the interests of SOEs. This policy, however, was biased and only brought a short-term success in winning the blue sky. As our empirical data shows, while the degree of PM2.5 in the BTH region was significantly reduced at the beginning of the campaign, the record of achievement could not be maintained. The biased direction of air quality governance caused problems in other areas. Placing the focus of reducing PM2.5 on dumping coal burners led to insufficient heating for many households in the northern provinces during the winter. This reveals that the original plan was not sustainable, and the direction of air quality governance eventually needed to be adjusted. Accordingly, while China's authoritarian leaders may have a firm determination to improve the air quality, the strategy of defecting responsibility for air pollution represents a strong bias of the policymaking direction.

Therefore, the shortcomings of Chinese-style authoritarian environmentalism are not only the possible non-cooperation of local officials during the policy implementation process but also the problems in the policy formulation process. In order to maintain regime stability, the authoritarian government cannot completely exclude the voice of business sector. It still needs to take into consideration the interest of state-owned enterprises because keeping the developmentalist goal is critical to the regime survival.

At the same time, although authoritarian environmentalism argues that excluding social voices helps avoid interference with environmental protection policies, the result of excluding ordinary people from the policymaking process is that the authoritarian government cannot obtain adequate grassroots information. The lack of a feedback mechanism has hindered the improvement of overall governance quality in China across different quality-of-life issues. As our analysis of the "Coal-to-Gas Switch" campaign indicates, the Chinese government had to put the brakes on the policy when abandoning small coal stoves caused insufficient heating and an extremely freezing winter. In other words, the authoritarian government only realized that the original direction of governance must be changed after the adverse consequence occurred. Compared to democratic regimes where grassroots information is relatively easy to be obtained through free media reports and frequent public opinion polls, authoritarian governments do not have such tools to collect this information. In democracies, when a policy causes bad effects at the grassroots level, the government can make adjustments in a quicker way according to societal feedbacks. In authoritarian regimes, however, without adequate channels to know about the policy implementation, the government can only rely on reports made by local governments or by the inspectors sent out by the central government. In many cases, local governments in an authoritarian regime will not tell the truth by themselves and central inspection is usually inaccurate because local governments are inclined to hide information. The central government hence does not have immediate information about the quality of its policy, and can only adjust the bad policy when the implementation has caused severe consequences. Accordingly, the cost for an authoritarian government to adjust its policy direction is substantially high. The Chinese government is not as adaptable as Eric X. Li claimed in his 2013 TED Talk (Li, 2013). Only when severely negative effects are produced during policy implementation will the Chinese government realize the need for policy adjustment. This is the fundamental flaw of the authoritarian environmentalism model.

Notes

1 The twenty-six prefecture-level cities are Shijiazhuang, Tangshan, Langfang, Baoding, Cangzhou, Hengshui, Xingtai, and Handan (Hebei Province); Taiyuan, Yangquan, Changzhi, and Jincheng (Shanxi Province); Jinan, Zibo, Jining, Dezhou, Liaocheng, Binzhou, and Heze (Shandong Province); and Zhengzhou, Kaifeng, Anyang, Hebi, Xinxiang, Jiaozuo, and Puyang (Henan Province).
2 These enterprises include non-ferrous metal smelting processing, rubber production, tanning, chemical industry, ceramic firing, casting, wire mesh processing, steel rolling, refractory materials, carbon production, lime kiln, brick kiln, cement grinding station, waste plastic processing, and small manufacturing and processing enterprises such as printing and furniture that use inks, adhesives, organic solvents, etc.
3 Please see Figure 8.2 for the comparison of the PM2.5 level in winter and in summer.

4 Energy Foundation China is a charitable organization supervised by the National Development and Reform Commission. For details of the Foundation, please see the official website at www.efchina.org/About-Us-en/Introduction-en

5 All the original PM2.5 data is from the "Real-time Air Quality Report of All Chinese Cities," a dataset supervised by China's National Environmental Monitoring Centre, an institute under the Ministry of Ecology and Environment. The Dataverse of the Center for Geographic Analysis, Harvard University has compiled the data from 2015 through 2016 in a format that is ready for analysis (Berman, 2017). For the PM2.5 data from 2017 through 2019 we directly downloaded it from the Real-time dataset (Real-time Air Quality Report of All Chinese Cities, 2017–2019).

6 The number of local monitoring stations has been increased every year. In 2016, there were 1,497 ground stations across the country. In 2017, the number was increased to 1,563 stations. In 2018, there were 1,605 in total and in 2019, there were 1,641 stations. We calculate the PM2.5 value of Chinese cities from all these known stations.

References

Ai, M. (2016, October 19). North China is Plagued by Heavy Smog Again. *Radio France Internationale Taiwan*. www.rfi.fr/tw/%E4%B8%AD%E5%9C%8B/20161019-%E8%8F%AF%E5%8C%97%E5%9C%B0%E5%8D%80%E5%86%8D%E6%AC%A1%E5%8F%97%E5%88%B0%E9%87%8D%E5%BA%A6%E9%9C%A7%E9%9C%BE%E5%9B%B0%E6%93%BE

Beeson, M. (2010). The Coming of Environmental Authoritarianism. *Environmental Politics*, 19(2), 276–294.

Beijing Municipal Commission of Development and Reform. (2016, August 10). *Notice on Increasing the Support of the Policy of Switching Coal to Clean Energy*. http://zfs.mee.gov.cn/hjjj/gjfbdjjzc/hjczzc1/201608/t20160810_362121.shtml

Berman, L. (2017). National AQI Observations (2014–05 to 2016–12). *Harvard Dataverse*, V3. https://doi.org/10.7910/DVN/QDX6L8

Bo, X., Xu, J., Du, X., Guo, Q., Zhen, R., Tian, J., Cai, B., Wang, L., Ma, F., & Zhou, B. (2017). Impacts Assessment of Steel Plants on Air Quality over Beijing-Tianjin-Hebei Area. *China Environmental Science*, 37, 1684.

Chen, G. C., & Lees, C. (2018). The New, Green, Urbanization in China: Between Authoritarian Environmentalism and Decentralization. *Chinese Political Science Review*, 3, 212–231. https://doi.org/10.1007/s41111-018-0095-1

Cheng, G. Y. (2017, November 9). Shanxi Province 2017–2018 Comprehensive Air Pollution Governance and Action Plan for the Autumn and Winter. *Shanxi Daily*. http://sx.people.com.cn/GB/n2/2017/1109/c189130-30902393.html

Dong, R. (2018, January 27). The Strom of Environmental Protection Supervision Sweeps Beijing, Tianjin, Hebei and the Surrounding Areas. *The Economic Observer*. https://finance.sina.cn/2018-01-27/detail-ifyqyqni3673199.d.html?wm=3049_0015

Eaton, S., & Kostka, G. (2014). Authoritarian Environmentalism Undermined? Local Leaders' Time Horizons and Environmental Policy Implementation in China. *The China Quarterly*, 218, 359–380. https://doi.org/10.1017/S0305741014000356

Energy Foundation China. (2016, January). *One Article to Understand the Causes, Hazards and Solutions of China's Smog*. www.efchina.org/News-zh/Program-Updates-zh/RestoringBlueSkies-zh/pop_science/pop_science_20160122

Gao, J. (2017, May 18). The Largest Environmental Protection Inspection in History Has Been Launched for More Than a Month: What Problems Have Been Found? Tracing the Environmental Protection Supervision in 28 Cities of the Beijing-Tianjin-Hebei Region and Its Surrounding Areas. *Xinhua News Agency*. www.gov.cn/xinwen/2017-05/18/content_5195092.htm

Gilley, B. (2012) Authoritarian Environmentalism and China's Response to Climate Change. *Environmental Politics*, 21(2), 287–307, https://doi.org/10.1080/096 44016.2012.651904

Greenpeace East Asia. (2018, January 12). Analysis of Air Quality Trends in 2017. *Greenpeace*. www.greenpeace.org/static/planet4-eastasia-stateless/2019/11/ 2aad5961-2aad5961-analysis-of-air-quality-trends-in-2017.pdf

Holdaway, J., Yang, L., Li, Y., Wang, W., Vardoulakis, S., & Krafft, T. (2018). Taking Action on Air Pollution Control in the Beijing-Tianjin-Hebei (BTH) Region: Progress, Challenges and Opportunities. *International Journal of Environmental Research and Public Health*, 15(2), 306.

Huang, Y., & Yang, D. (2002). Bureaucratic Capacity and State-society Relations in China. *Journal of Chinese Political Science*, 7(1), 19–46. https://doi.org/10.1007/ BF02876928

Ji, W. (2017, April 29). The Battle for Replacing Scattered Coal Has Started. *China Electric Power News*, 010. https://ie.cnki.net/kcms/detail/detail.aspx?filename= CDLB201704290100&dbcode=XHDN_XNYN&dbname=XNYTLKCCNDLAS T2017&v=

Kostka, G., & Nahm, J. (2017, September). Central-local Relations: Recentralization and Environmental Governance in China. *The China Quarterly*, 231, 567–582. https://doi.org/10.1017/S0305741017001011

Li, E. X. (2013, June). A Tale of Two Political Systems. *TED Global 2013 Talk*. www.ted.com/talks/eric_x_li_a_tale_of_two_political_systems?language=en

Li, J., & Feng, X. (2017, December 19). Collectively Promoting the "Coal to Gas" Switch is a Project of Morality and Good Governance. *China Environment News*, 003. http://cnki.sjzkz.com/KCMS/detail/detail.aspx?filename=CHJB20171219 0030&dbcode=CCND&dbname=CCND2017

Li, X., Yang, X., Wei, Q., & Zhang, B. (2019, June). Authoritarian Environmentalism and Environmental Policy Implementation in China. *Resources, Conservation and Recycling*, 145, 86–93.

Lieberthal, K. G., & Lampton, D. M. (1992). *Bureaucracy, Politics, and Decision Making in Post-Mao China*. Berkeley, CA: University of California Press.

Lieberthal, K. G., & Oksenberg, M. C. (1988). *Policy Making in China: Leaders, Structures, and Processes*. Princeton, NJ: Princeton University Press.

Livingston, S. (2020). *The Chinese Communist Party Targets the Private Sector*. Center for Strategic and International Studies (CSIS). www.csis.org/analysis/ chinese-communist-party-targets-private-sector.

Ministry of Environmental Protection of the People's Republic of China. (2016a). *Technical Guidelines for Comprehensively Governing the Use of Scattered Coal and Burning Pollution in Rural Areas*. www.mee.gov.cn/gkml/hbb/bgth/201603/ W020160315372078414744.pdf

Ministry of Environmental Protection of the People's Republic of China. (2016b, October). *Technical Guidelines for Comprehensively Governing the Civil Use of Coal Burning Pollution*. www.mee.gov.cn/gkml/hbb/bgg/201610/W02016 1031388726871449.pdf

Ministry of Environmental Protection of the People's Republic of China, The Government of Beijing City, The Government of Tianjin City, & The Government of Hebei Province. (2016, June 20). *Strengthening Measures for the Prevention and Control of Air Pollution in the Beijing-Tianjin-Hebei Region (2016–2017)*. www.coolsorption.com.cn/wap/content/?53.html

Ministry of Environmental Protection of the People's Republic of China, National Development and Reform Commission, Ministry of Finance, National Energy Administration, The Government of Beijing City, The Government of Tianjin City, The Government of Hebei Province, The Government of Shanxi Province, The Government of Shandong Province, & The Government of Henan Province. (2017a, February 17). *Notice on Issuing the Work Plan for the Prevention and Control of Air Pollution in the Beijing-Tianjin-Hebei Region and Its Surrounding Areas in 2017*. http://dqhj.mee.gov.cn/dtxx/201703/t20170323_408663.shtml

Ministry of Environmental Protection of the People's Republic of China, National Development and Reform Commission, Ministry of Industry and Information Technology, Ministry of Public Security, Ministry of Finance, Ministry of Housing and Urban-Rural Development, Ministry of Transport, State Administration for Industry and Commerce, General Administration of Quality Supervision, Inspection and Quarantine, National Energy Administration, The Government of Beijing City, The Government of Tianjin City, The Government of Hebei Province, The Government of Shanxi Province, The Government of Shandong Province, & The Government of Henan Province. (2017b, August 21). *2017–2018 Comprehensive Air Pollution Governance and Action Plan for the Autumn and Winter in the Beijing-Tianjin-Hebei Region and its Surrounding Areas*. www.mee.gov.cn/gkml/hbb/bwj/201708/t20170824_420330.htm

Modern Express. (2013, December 8). *Experts Talk About Smog Management: Jiangsu Has a Large Proportion of Heavy Industry Which Makes the Transformation Difficult*. https://js.qq.com/a/20131208/001050.htm

National Development and Reform Commission of the People's Republic of China, National Energy Administration, Ministry of Finance, Ministry of Environmental Protection, Ministry of Housing and Urban-Rural Development, State-owned Assets Supervision and Administration Commission, General Administration of Quality Supervision, Inspection and Quarantine, China Banking Regulatory Commission, China Securities Regulatory Commission, & Logistic Support Department of the Central Military Commission. (2017, December 20). *Clean Winter Heating Plan in Northern China (2017–2021)*. www.gov.cn/xinwen/2017-12/20/content_5248855.htm

Real-time Air Quality Report of All Chinese Cities. (2017–2019). *China's National Environmental Monitoring Centre*, Ministry of Ecology and Environment. www.cnemc.cn/sssj/

State Council of the People's Republic of China. (2013, September 12). *Notice of the State Council on Issuing the Air Pollution Prevention and Control Action Plan*. www.gov.cn/zwgk/2013-09/12/content_2486773.htm

State Council of the People's Republic of China. (2016, December 5). *Notice of the State Council on Issuing the "13th Five-Year" Ecological Environment Protection Plan*. www.gov.cn/zhengce/content/2016-12/05/content_5143290.htm

Sutter, K. M., & Sutherland, M. D. (2021). China's Economy: Current Trends and Issues. *Congressional Research Service Report#: IF11667*. https://crsreports.congress.gov/product/details?prodcode=IF11667

Tang, Q., Lei, Y., Yan, G., Xue, W., & Wang, X. (2020). Characteristics of Heavy Air Pollution in Beijing-Tianjin-Hebei and the Surrounding Areas During Autumn and Winter and Policy Recommendations. *IOP Conference Series: Earth and Environmental Science*, 569, 012041.

Tangshan Statistical Bureau. (2017, May 29). *Statistical Communiqué on the 2016 National Economic and Social Development of Tangshan*. www.cnstats.org/tjgb/201705/hbtss-2016-ibj_4.html

Wells, P. (2007). The Green Junta: Or, Is Democracy Sustainable? *International Journal of Environment and Sustainable Development*, 6(2), 208–220.

Wen, J. (2016, March 7). Hebei Has Invested 24 Billion Yuan in Governance: The Atmosphere of Beijing, Tianjin and Hebei will Improve Significantly in 3 to 5 years. *Beijing Times*. http://politics.people.com.cn/n1/2016/0307/c1001-28176802.html

Wintrobe, R. (1998). *The Political Economy of Dictatorship*. Cambridge: Cambridge University Press.

Wintrobe, R. (2001). How to Understand, and Deal with Dictatorship: An Economist's View. *Economics of Governance*, 2(1), 35–58. https://doi.org/10.1007/s10101-001-8001-x

Wu, X., Yang, D. L., & Chen, L. (2017). The Politics of Quality-of-Life Issues: Food Safety and Political Trust in China. *Journal of Contemporary China*, 26(106), 601–615. https://doi.org/10.1080/10670564.2017.1274827

Xu, X. (2016, October 18). Governing the Air Pollution Must Put into Practice the Concept of Shared Responsibility. *Lanzhou Daily*, 003. http://gansu.gansudaily.com.cn/system/2016/10/18/016460499.shtml

Ye, B. (2016, October, 14). The Beijing-Tianjin-Hebei Region is Again in Heavy Pollution. *Voice of America Cantonese*. www.voacantonese.com/(X(1)S(hwslsl41luwkiccludogvegd))/a/haze-beijing-20161014/3550397.html

Zhu, Y. (2011). "Performance Legitimacy" and China's Political Adaptation Strategy. *Journal of Chinese Political Science*, 16, 123–140.

9 Social history of air pollution in Japan

Focusing on the second stage caused by automobiles

Manami Horihata

9.1 Introduction

This chapter focuses on the Tokyo Air Pollution Lawsuit, which concerned the treatment of pollution and health hazards after Japan instituted new air pollution control measures. The Tokyo Air Pollution Lawsuit was filed as part of an effort to establish a new certification after legal revisions ended the pollution patient certification.

The proceedings admitted that Tokyo, which manages the road, was responsible for causing air pollution. In 1973, the environmental standard value for nitrogen dioxide, the causative agent of asthma, was set at an average of 0.02 ppm per day. Around the same time, the enactment of the Musky Law was debated in the USA, where air pollution by automobiles became a social problem. In advance discussions, a major American automaker argued that the standards were too strict, so loose standards were set. In 1978, Japan relaxed the standard value of nitrogen dioxide to "within 0.04 to 0.06 ppm on average per day" in line with the US Musky Act. Scientists objected to the change, claiming it had no scientific basis, but the country forced a relaxation of the standards.

Japan thus sought to address Sulfur oxides with control technology and nitrogen dioxide by relaxing environmental standards.

The discontinuation of the air pollution patient certification meant that companies that had implemented pollution control measures had already fulfilled their responsibilities. As cars replace new sources of pollution, there is no relief under the Pollution and Health Damage Compensation Act. On the other hand, whether the cause of air pollution is soot from factories or automobile exhaust gas, there is no difference in living in dirty air, and respiratory diseases develop by inhaling dirty air for a certain period of time . . . Air-polluted patients whose onset timing was after March 1, 1988, are no longer relieved by the revision of the law.

Since Japan operates a universal health insurance system, asthma patients diagnosed after the discontinuation of certification receive a specific insurance subsidy—from, for example, the National Health Insurance—and must pay their own expenses. Although the disease remained the same, jurisdiction over it shifted from the Ministry of the Environment to the Ministry of Health, Labor and Welfare. In medicine, asthma is treated as a "common" type of illness referred to as a lifestyle-related disease.

DOI: 10.4324/9781003211747-12

These developments highlight the risk of the politics of pollution in Japan. The nation sought to leave pollution patients behind and obscure the responsibility of the perpetrators. Indeed, Japan's pollution control measures have often prioritized corporate interests, though this may not be obvious at first glance.

9.2 Pollution damage in Japan from a structural perspective

In "Damage Structure Theory," the environmental sociologist Nobuko Iijima analyzed Japan's pollution problems, especially Minamata disease in Kumamoto (Iijima, 1993: 83).

Using this analysis, it can be seen that patients with air pollution suffer from damages such as daily life, family relationships, employment and household burdens, and adverse effects on cultural behavior. Illness also deprives patients of leisure activities and other sources of enjoyment and makes them feel alienated by, for example, differentiating them from healthy people and leaving them unable to contribute to society. Moreover, illness can cause mental as well as physical damage, including making people incapable of comprehending their surroundings. Indeed, prolonged illnesses often worsen families' relationships and the situations around them. Patients and their families are frequently forced to change their living plans and experience a reduced standard of living.

Five factors determine the depth of damage caused by illness: (1) degree of health damage; (2) the role and status of the person whose health is damaged in the home; (3) the social status/social class of the victim or his/her family; (4) the victim's group affiliations; and (5) external factors. The fifth factor includes perpetrators, the government, medical personnel, scholars, the general public, and mass media (Iijima, 1993: 84). Very importantly, damage from pollution is likely to be latent.

By not complaining to others about the difficulties they face, pollution patients often hid their illnesses. The latent nature of the damage makes it easy to claim that no damage occurred; indeed, until the early 1960s, victims of pollution and occupational accidents received no compensation or relief for the considerable damage they experienced (Iijima, 1993: 79). At that time, many of the victims of serious pollution were low-ranking workers, farmers, fishermen, or consumers who lived extensively, unrelated to local or social activities. Therefore, forming systematic protest and resistance movements proved difficult, and those that did form experienced constant pressure (Iijima, 1993: 79).

Health hazards caused by pollution affect people who lack physical strength more severely. The effects are most pronounced in infants and the elderly, followed by pregnant women, women, and men. Since the male population has the greatest political influence, addressing the harm they experience is often difficult for women. In addition, even in the male group, the severity of harm varies based on social class and is often underestimated, making it difficult for victims to achieve recognition (Iijima et al., 2007).

Although it does not cause serious health damage such as pollution, it is a typical example where future health damage can be imagined and a nuclear disaster that leads to an uneasy daily life can be analyzed by damage structure theory. Fujikawa, who studies nuclear disasters, claims that the underestimation of damage involves dividing victims based on disparities in compensation and relief (Fujikawa, 2018: ii). He points out that this acts in combination with incomprehension, discrimination against victims, and the victim's own giving up to cause forgetfulness and ignorance of damage in society (Fujikawa, 2018: ii).[1] Although the air quality has become cleaner than it was during the period of severe pollution, urban patients with respiratory diseases, as well as victims of nuclear disasters, are experiencing a disregard for their fears, an underestimation of the damage, and an ambiguity of responsibility for the harm caused.

9.3 Historical development of air pollution

9.3.1 From pollution asthma to common asthma

The Japanese people have experienced four major pollution-related diseases: Kumamoto Minamata disease, Niigata Minamata disease, Itai-itai disease, and Yokkaichi pollution asthma. Of these, Yokkaichi pollution asthma, caused by sulfur oxides and nitrogen oxides emitted from oil refineries, has become a symbolic example of air pollution.

In 1969, Japan's parliament enacted time-limited legislation, "Special Measures Law Concerning Relief of Health Damage Related to Pollution," and patient relief efforts began. In 1970, the extraordinary parliament enacted 14 pollution-related laws including the Basic Law on Pollution Control. In 1974, the Permanent Law, "Pollution and Health Damage Compensation Law," was enacted separately for regions and designated diseases. Under the law, areas with severe air pollution are designated as Class I areas, and compensation is provided to patients who develop respiratory diseases in these areas.

After that, the combined efforts of the public and private sectors culminated in the development of technology to remove the sulfur oxides contained in petroleum and coal, which is a cause of asthma, first from fuel and then from combustion gas. In February 1988, improvements in air quality led the government to revise the Pollution and Health Damage Compensation Law to cancel the new certification that enabled air pollution patients to receive compensation for health damage, including medical expenses and the costs of medicine. This amendment meant that the law only continued to cover about 51,000 already-certified people.

Regarding the revision to the law, the Environmental Restoration and Conservation Agency, which oversees pollution and health damage prevention, issued the following explanation:

> The current situation of air pollution is different from the situation of significant air pollution in the 1950s and 1940s, and it cannot be said that it is the main cause of illnesses such as asthma. Not so much, but it's possible that they have some effect on these illnesses.[2]

The government claimed that air pollution had been reduced, and it could no longer be deemed the main cause of asthma. Many local governments, which are designated as Type-1 areas, opposed the revision of the law, but the government forced it through. Since March 1, 1988, the national government has decided to focus on the prevention of health hazards for people living in areas with air pollution, including the implementation of pollution health damage prevention projects, so that air pollution patients who used to be helped are no longer helped.

As a result of this revision of the law, the government created a new 50-billion-yen fund for preventative projects called the "Pollution and Health Damage Prevention Fund." The fund included 40 billion yen from fixed sources, comprising the direct causers of air pollution, 5 billion yen from automobile manufacturers engaged in activities related to air pollution, and 5 billion yen from the national government. The investment income from this fund was to be used to carry out preventive projects, which are projects that ensure and restore health and projects that contribute to improving the environment.

Due to a revision of the law, after March 1988, patients who newly developed asthma, even in the former Type 1 designated areas, had to pay their own medical expenses through the National Health Insurance system.[3]

In the case of pollution asthma, the pollution cause was responsible for the entire amount, but such cases did not fall under this framework and general medical resources covered the expenses. This revision to the law removed asthma from the "pollution illness" list and recategorized it as a "general illness," as the medical expenses financial resources reveal. According to a patient survey conducted by the Ministry of Health, Labor, and Welfare in 2017, Japan is home to 1,117,000 asthma patients. In the list of diseases such as hypertension (9937,000 people) and periodontal disease (3983,000 people), asthma ranks 9th.[4] Asthma has thus come to be treated as a "common" disease rather than a pollution disease.

9.3.2   From Yokkaichi pollution proceedings to Tokyo air pollution proceedings

Among the four major pollution-related diseases, Yokkaichi pollution asthma has made the air pollution problem in areas where oil refineries are concentrated widely known. At the time of the Yokkaichi pollution trail, the petroleum complex along the coast where oil tankers were likely to berth generated significant air pollution, meaning pollution areas were concentrated in the bay.[5]

After the Yokkaichi pollution trail, the government enforced emission restrictions on factories, a fixed emission source, and improvements in emission control technology reduced sulfur oxides—the main cause of asthma at that time. As these regulations took effect during the 1970s and 1980s, automobiles emerged as the main source of air pollution.

Developments in air pollution lawsuits confirm this change; after the Yokkaichi pollution lawsuit, in which the factories were defendants, roads and automobile manufacturers became the primary defendants in such trials (Table 9.1).

Table 9.1 Air pollution proceedings after Yokkaichi pollution trial

	plaintiff	defendant	From primary complaint to settlement
Chiba Kawasaki Steel Pollution Lawsuit	Approved persons and their families, citizens living or commuting around Kawasaki Works	Factory (Kawasaki Steel)	1975~1992
Nishiyodogawa Air Pollution Lawsuit	Patients living or commuting in Nishiyodo Ward who are certified by the Public Health Act as having pollution	Factories (10 companies), roads (Country and Hanshin Expressway Public Corporation)	1978–1995 (company) 1998 (National Highway Authority)
Kawasaki Pollution Lawsuit	Patients who are living or commuting in Kawasaki City and certified by the Public Health Act	Factories (12 companies), roads (Country and Metropolitan Expressway Public Corporation)	1982–1996 (Company) 1999 (National Highway Authority)
Kurashiki Pollution Lawsuit	Patients who are living or commuting in Kurashiki City and are certified by the Public Health Law	Factories (8 companies)	1983~1996
Amagasaki Air Pollution Lawsuit	Patients living in southern Amagasaki or commuting to work under the Public Health Law who have been certified as having an illness and their survivors	Factories (9 companies), roads (Country and Hanshin Expressway Public Corporation)	1988–1999 (Company) 2000 (National Road Authority)
Nagoya Southern Air Pollution Lawsuit	Nagoya City Tokai City and its surroundings who live or commute to the public health law under the Public Health Law and their survivors	Factories (11 companies), roads (Country)	1989~2001
Tokyo Air Pollution Lawsuit	Patients or survivors who have bronchial asthma, chronic bronchitis, or emphysema while living or commuting in Tokyo (including those who have not been certified by the Public Health Act)	Diesel automobile manufacturers (7 companies), roads (Country, Tokyo, Metropolitan Expressway Public Corporation)	1996~2007

Note: Excerpted from the website of the Environmental Restoration and Conservation Agency "Air Pollution and Trials in Records." www.erca.go.jp/yobou/saiban/

The lawsuits that have highlighted road pollution as a social problem include the Amagasaki air pollution lawsuit, the Kawasaki pollution lawsuit, the southern Nagoya air pollution lawsuit, and the Tokyo Air Pollution Lawsuit, though the latter also addressed air pollution in an industrial area on the Pacific coast. Of these lawsuits, only Tokyo Air Pollution Lawsuit featured an automobile manufacturer rather than a factory as the defendant.

Japan enacted the "Pollution and Health Damage Compensation Law" in 1973 and was immediately hit by the oil shock. Then, in July 1978, the government relaxed the environmental standards for nitrogen oxides by two to three times in order to stimulate economic activity. This was because motorization had greatly increased in Japan, and automobile exhaust emissions had replaced other important issues for the government. The establishment of environmental standard values for nitrogen oxides was based on the Muskie Act in the USA. However, the Muskie Act in the USA, which was used as a reference, was a relaxation of the nitrogen oxide emission standards that had been presented to automakers in advance, as they could not meet the standards. This relaxation was also criticized by scientists, as it was not decided after considering the effects on the human body.

This relaxation was also criticized by scientists, as it was not decided after considering the effects on the human body. Masashichi Nishio, a physician, pointed out the inadequacy of the epidemiological study by the expert committee that determined these figures, saying that it did not take into account the population groups, even though there was a system in which elderly people were more likely to develop the disease (Nishio, 1979). In fact, in its 1979 Environmental White Paper states, the Environmental Agency stated: "it is inappropriate to use epidemiological data as an element of comprehensive judgment, or that statistical analysis focusing only on epidemiological data has an annual average of 0.02 to 0.03 ppm." It is stated that there are criticisms regarding the adoption of epidemiological data that "cannot be derived."[6]

In the next section, we examine the 1996 Tokyo Air Pollution Lawsuit, which highlighted the problem of pollution caused by roads.

9.4 Tokyo air pollution lawsuit

9.4.1 Proceedings summary

Filed in 1996, the Tokyo Air Pollution Lawsuit included 188 uncertified patients who did not receive compensation for medical treatment from the Pollution and Health Damage Compensation System as plaintiffs. In this case, residents of 23 wards in Tokyo demanded damages from and an injunction against the defendants to compensate them for the suffering they experienced after contracting respiratory diseases caused by automobile exhaust emissions. The defendants included the Tokyo Metropolitan Government road managers, the government, the Metropolitan Expressway Public Corporation,[7] and automobile manufacturers.

The proceedings consisted of (1) the inclusion of an automobile manufacturer, (2) the problem of defining area pollution because the road network was stretched like a mesh, and (3) the inclusion of uncertified patients.

With regard to (1), this is because the manufacturer's liability of automobile manufacturers had never been questioned in Japan before. Regarding (2), so-called road pollution, such as air pollution, noise, and vibration, has been considered to occur in a linear fashion along roads. As roads were built, it was thought that air pollution would occur on an area-wide basis, as the lines overlapped with each other. As for (3), patients who once developed the four air pollution-designated diseases after their certification was terminated have joined the court case to seek compensation.

On October 29, 2002, the first instance judgment was handed down, denying the plaintiffs' claim of surface contamination, but acknowledging that some of the plaintiffs had actually suffered damage. The ruling acknowledged the responsibility of the road administrator, the national government, the Tokyo Metropolitan Government, and the Metropolitan Expressway Public Corporation. The court rejected the notion that the automakers were responsible, but stated that they had a social responsibility to manufacture and sell automobiles with less environmental impact.

The trial court also pointed out that, at the latest around 1973, it was "foreseeable" that a huge number of automobiles would be concentrated and clustered in Tokyo, that local pollution from automobile exhaust would be likely to occur in areas served by major roads, and that residents along the roads might suffer from respiratory diseases caused by exhaust gas (Yoshimura, 2003). On August 8, 2007, the defendants accepted the settlement recommendation presented by the court. The settlement required the defendants to invest about 20 billion yen over five years to develop a new medical assistance system and strengthen pollution control measures. Of this 20-billion-yen total, 6 billion yen came from the government, paid from the Pollution and Health Damage Prevention Fund operated by the Environmental Restoration and Conservation Agency.

Since the settlement stipulated that the medical voucher benefits would be reviewed after five years, the Tokyo Pollution Patients and their Families Association lobbied the Tokyo Metropolitan Government, the national government, and the automakers to ensure that the medical voucher benefits would continue, but both the automakers and the national government later refused to bear the additional burden. Due to a lack of financial resources, the Tokyo Metropolitan Government discontinued new certification for medical expense subsidies beginning in FY2014. In 2015, the maximum number of certified patients under this system was 97,874. In FY2019, the number of newly certified patients had decreased to 59,905 (Tokyototaikioseniryouhijyoseikentouiinnkai, 2008).[8]

On February 18, 2019, 90 former plaintiffs in the Tokyo Air Pollution Lawsuit filed for mediation with the National Environmental Dispute Coordination Commission, seeking to create a new system that would fully subsidize medical expenses. But in 2021, on December 28, mediation was unsuccessful.

9.4.2 Damage revealed in lawsuit

Part 8, Chapter 1, "Introduction to Damage" of the "Final Preparation Document" from the Tokyo Air Pollution Lawsuit defines eight types of damage: (1) Physical damage (disease pain, intractability, side effects of drugs, fear of death); (2) Destruction of daily life (sleeping, walking, change of clothes, eating, bathing, excretion, housework, etc.); (3) Destruction of family life (burden on family members such as nursing, impact on family members' work/school, loss of family group, conflict between families, mental distress of family, difficulty having a family); (4) Economic damage (disadvantages in work such as job change, retirement, decrease in revenue, increase in expenses such as medical expenses, relocation medical treatment, purchase of goods, home remodeling, going to hospital, difficulty in employment, loss of work worth); (5) Destruction of social and cultural life (destruction of friendship, etc.); (6) Difficulties in school life (bullying, academic difficulties); (7) Mental damage (mental distress as illness, anxiety over possibility of seizures, sadness over lack of independence, dear of death, despair for life, lack of understanding of surroundings); and (8) Damage to uncertified patients (Yokemoto & Horihata, 2003).

The distinction between certified and uncertified is based on whether or not you are entitled to compensation for medical expenses, etc., and not on whether or not you are a pollution patient. Under the Pollution Health Damage Compensation Act, there was not only compensation for medical expenses, but also compensation for work depending on the severity of the illness. Testimony from patients in the Tokyo air pollution lawsuit confirms that they did not seek medical care in order to reduce medical costs, or reduce the prescribed dosage in order to save money on medication. It has also been confirmed that some patients have been suffering from asthma so severe that they are unable to continue working, and that family relationships have deteriorated to the point that the family has been separated and is now living on welfare(Yokemoto & Horihata, 2003).[9]

9.4.3 Actual damage described in questionnaire responses

In August 2008, as a result of the settlements, the Tokyo Metropolitan Government established a medical expenses subsidy system and revised and expanded the ordinance enacted in 1972 to cover all ages, including patients with bronchial asthma over 18.[10]

From May 2009 to the end of August 2009, about 3,000 sheets were distributed and 652 forms were collected. A total of 78,000 asthma patients were covered, and as of 2010, more than 50,000 asthma patients have been newly certified under the revised ordinance.

Since privacy concerns prevented us from obtaining the patient list from Tokyo, we distributed the questionnaire with the cooperation of the Tokyo Democratic Medical Institutions Association and the Tokyo Pollution Patients and Family Association and collected them using the self-administered mail method. The survey period was set at 3 months because the interval between visits varies from patient to patient.

First, regarding respondent attributes, the area of residence, gender, age group, and household income warrant mention. Areas not included in the former designated area were also subject to the ordinance, so about 50% of the people lived outside the former designated area. This means the spread of air pollution was caused by automobile exhaust emissions, not by factories (Ozaki et al., 2011).

Forty percent of the respondents were male and 60% were female. By age group, 9.0% were 19–39 years old, 17.0% were 40–59 years old, 49.2% were 60–74 years old, and 24.7% were 75 years old or older. People in their 60s and above thus accounted for more than 70% of the sample.

The influence of age was reflected in household income. Household incomes of less than 1 million yen accounted for 8.1% of the sample, 1 to 2.99 million yen for 39.3%, 3 to 5.99 million yen for 34.8%, and 6 million yen or more for 12.3%. About 50% of all respondents had total household incomes less than 3 million yen. 17.7% of those under the age of 40 and 26.4% of those aged 40–59 had annual household incomes of less than 3 million yen, indicating that health hazards are concentrated among the economically vulnerable.[11] Next, we present the characteristic results.

In the questionnaire, we asked about the decreases in medical expenses and drug costs patients experienced after receiving the new medical vouchers. The average decrease in medical expenses was 2374 yen, and the average decrease in drug costs was 3483 yen.

Of the patients who had an attack at least once a month or were hospitalized at least once a year, about 20% were still prescribed seizure arrest and had severe medical conditions.

Table 9.2 provides an overview (by age group) of patients' negative experiences—including putting off medical examinations, saving medication, and income reductions—before they received the medical vouchers.

Table 9.2 Negative experience caused by asthma before receiving medical tickets (by age group) n=652 unit %

	Refrain from medical treatment	*Drug saving*	*Income decrease*	*Unemployment*	*Life design transformation*
19–39 years old	39.0	39.0	27.1	13.2	23.7
40–59 years old	40.5	39.6	24.3	8.6	17.1
60–74 years old	28.3	20.6	22.4	6.2	18.4
75 years old and over	17.4	5.0	12.4	1.3	11.8
all ages	28.7	21.6	20.7	7.4	17.0

※ Prepared by the author based on survey results.

Table 9.3 Negative experience caused by asthma before receiving medical tickets (by household income) n=652 unit %

	Refrain from medical treatment	Drug saving	Income decrease	Unemployment	Life design transformation
Less than 1 million yen	43.4	9.4	20.8	13.2	26.4
1 to 2.99 million yen	32.0	25.0	24.2	8.6	17.6
3–5.99 million yen	24.7	23.3	20.3	6.2	15.9
6 million yen or more	21.3	18.8	15.0	1.3	11.3
No answer	25.0	11.1	11.1	11.1	19.4
all	28.7	21.6	20.7	7.4	17.0

※ Prepared by the author based on survey results.

In 2003, the national government decided that 30% of medical expenses and medicine costs should be borne by those under 70 years old and 10% should be borne by those aged 75 and over. This has decreased the percentage of people aged 75 and over who have refrained from medical treatment, and meant that, in general, the younger people are, the more negative their experiences.

Next, in Table 9.3, we look at negative experiences by household income. The lower the income people had, the more they refrained from medical consultations. Meanwhile, the life design change results reveal that the response rate increases as annual household income decreases.

These results confirm some of the related damages pointed out by Iijima

Next, Table 9.4 shows the top-10 positive effects of medical vouchers.

The new medical vouchers were introduced as a result of the air pollution patients' legal victory. Thus, examining the effects of medical vouchers helps determine whether the negative impacts of air pollutants have been financially and mentally alleviated.

Table 9.4 lists the top-10 items. Respondents provided multiple answers to this question, so the total exceeds 100%.

The top-scoring item was "I can now be hospitalized and receive outpatient care without worrying about money," followed by "I'm glad that my illness was recognized as being due to pollution", and "I came to think that I would actively treat asthma." This result highlights the importance of addressing financial concerns. In addition, it shows that patients perceive medical vouchers as government acknowledgment that pollution caused their illnesses. In Japan, where the theory of personal responsibility for illness remains strong, such recognition saves patients mental anguish.

This medical voucher system was found to be of great benefit to air pollution patients. Based on the results, the program was to be reviewed every five years, but the patients who fought in the court demanded that the program be continued. However, the system could not be continued.

Table 9.4 Positive effect after receiving a medical ticket (%)

I can now be hospitalized and outpatient without worrying about money.	72.2
I'm glad that my illness was recognized as being due to pollution.	57.4
I came to think that I would actively treat asthma.	52.8
Asthma symptoms improved	36.3
I started to take the medicine as prescribed without saving the medicine I got from the hospital/pharmacy.	32.5
Not only asthma treatment, but also willingness to undergo tests	27.1
It became easier to make a request to the attending physician about the content of treatment.	26.8
My doctor has come to offer me various treatments without worrying about the burden of medical expenses.	25.5
Be positive in life	21.5
I'm confident that I can continue working.	14.6
Top 10 items listed n=652	

※ Prepared by the author based on survey results.

9.5   Japan's risk politics from the perspective of damage structure theory

9.5.1   *Underestimation of damage caused by "common" illnesses*

In general, people do not actively disclose their diseases. Air pollution patients can be controlled with medication to prevent attacks to some extent, so under normal conditions, people around them will not know that they are patients. Of course, patients certified by the Pollution and Health Damage Compensation Law are reluctant to tell others about their diseases (Horihata, 2009), and discrimination against pollution patients undoubtedly contributes to their silence in this regard.

The fact that pollution asthma is treated as normal asthma is significant for correcting discrimination, given the fact that pollution patients have been discriminated against. In Japan, there is a tendency to discriminate against people who are physically weak as being unable to lead a social life. Children with asthma have been considered to have a lack of physical strength as a personal factor (Horihata, 2009). There are many patients who would not suffer from asthma if the air were clean. The government does not recognize nitrogen oxides as a direct cause of health damage, but since nitrogen oxides are transformed into photochemical oxidants by ultraviolet rays and are also secondary particles of PM2.5, there is room for improvement in the environmental standard values.

Patients tend to blame themselves when they develop asthma and others do not. The survey results indicate that the distribution of new medical vouchers provided mental relief. In addition, the distribution of new medical vouchers eliminated the burden on households from patients who cannot work as much as they want, making it possible for them to live positively.

The Tokyo Metropolitan Government stopped new certifications in 2014, citing a lack of financial resources. Furthermore, in 2018, the government changed the system to provide medical subsidies only when the cost of medical care and medication is more than 6,000 yen. According to the results of the survey we took, the number of people eligible for medical subsidies will be reduced to about half; the number of medical vouchers as of 2019 is 59,905, which is a 40% decrease compared to the highest number of 97,874 in 2015. In addition, those under the age of 18 continue to receive full subsidies for medical expenses.

Patients who received new medical vouchers by 2014 and renew them can receive medical expense subsidies for costs beyond 6000 yen. However, since renewing takes time, nearly 40,000 people have stopped renewing.

Patients who developed the disease in FY2015 and those with medical bills of less than $6,000 were again faced with not only a financial burden, but also a negative experience compared to those who were medical ticket recipients and could receive compensation. The Tokyo Metropolitan Government does not release data, but as can be seen in the answers to the attributes of the questionnaire, many of the patients are elderly. The full amount of the National Pension Plan is 65,075 yen in FY2021, and 6,000 yen would represent 10% of the total amount. Even if a national pension recipient's medical expenses do not reach 6,000 yen, it will be a large percentage of their living expenses.

The media did not report that the new certification would be abolished in 2015 or that assistance would be limited to those who paid large amounts in 2018.[12] As a result, public opinion did not change, and the country remained largely indifferent. The fact that the media did not report on the underestimation of the damage also contributed to this indifference. Many automakers sponsor the media, contributing to a structure that defends perpetrators in such cases. Iijima cites an external factor as the determinant of the depth of damage (5), but it seems that the system itself is fostering indifference.

9.5.2 The end of the line: funding exhaustion

Although many air pollution patients benefited from receiving medical vouchers as a result of the Tokyo Air Pollution Lawsuit settlement, the funds were eventually exhausted.

The Tokyo Pollution Patients and Family Association envisioned this depletion of funds. It is usually reasonable to consider reducing the number of newly certified patients to ensure settlement-generated medical cost compensation continues. In these cases, however, the patient group hoped that many patients would be newly certified, and encouraged the Tokyo Metropolitan Government to disseminate posters and other information.

This is because the patient group believed that if the benefits of the new medical voucher were recognized, the national government would consider this system necessary and extend the measures implemented by the Tokyo Metropolitan Government nationwide.

When it was reported that the medical expense subsidies resulting from the settlement would begin, people involved in air pollution lawsuits such as Nishiyodogawa, Amagasaki, and Kawasaki complained and requested that the subsidies be expanded nationwide.[13] However, Tokyo continued to receive special treatment.

In the settlement, the government agreed to spend 6 billion yen from the principal (50 billion yen) of the Pollution and Health Damage Prevention Fund. This fund is a diversion because it was contributed to preventing pollution health damage on a regional basis following the abolition of the Pollution Health Damage Compensation Law in February 1988, which provided compensation to individuals. The government (Ministry of the Environment) refused to acknowledge the diversion and repeated its defense that the six billion yen was not a subsidy but part of a prevention project. This stance of the government has caused those involved in air pollution lawsuits outside of Tokyo, such as Nishiyodogawa, to grow distrustful of the government, saying, "Isn't Tokyo the only one being given preferential treatment?"[14]

The government explained the withdrawal from the Pollution and Health Damage Prevention Fund to Nippon Keidanren, to which the fund investor belongs, as follows:[15]

(1) The use of 6 billion yen is just a preventive business, not a medical expense subsidy as reported in some media reports. It is also specified in the settlement clause.

(2) Withdrawal of funds can be carried out based on Article 10 of the Supplementary Provisions of the Public Health Law, and there is no need to amend the laws and regulations accompanying this measure.

(3) 6 billion yen will be retired from the private contribution of the fund from the automobile industry, which is one of the defendants in the lawsuit.

(4) With the withdrawal of the fund, preventive project costs will decrease by about 200 million yen annually. While striving for even more efficient operation, we are currently requesting a new budget of 300 million yen, and we would like to use it for preventive project costs, so we would like to ask for the support of industry.

(5) We will not ask the industry to make new contributions to the prevention fund.

The government was worried that its decision to pay medical expenses would be viewed as political. The Tokyo Air Pollution Lawsuit involved special circumstances, including that the trial lasted 11 years and was ultimately decided by the Prime Minister. The government explained that it is unlikely that lawsuits will meet such high hurdles in the future, making additional withdrawals from the prevention fund less likely.

The Tokyo Metropolitan Government had been spending the fund directly without using the investment income from the fund, so the depletion of the fund was predictable.

In the 1988 amendment, the government succeeded in making the responsibility of the automobile manufacturers for the perpetration of the problem vague. They also succeeded in blending in the universal health insurance system, in which all citizens pay for their own health care, although in the case of pollution, the perpetrating companies pay all the medical costs.

The Tokyo Air Pollution Lawsuit settlement brought some clarity to the question of responsibility for harm.

However, the government's explanation to Keidanren shows that it planned to again obscure the issue.

The automakers responsible for the perpetration have a strong position in the industry. The voices of the victims do not reach them because the industry, such as Keidanren, is able to express its opinions at the national expert meetings. This is often the case with pollution. For example, when many local governments, which are designated as Type-1 areas, opposed the revision of the law, a fierce opposition movement was also waged in Yokkaichi City. Signatures against the amendment were collected from all over Japan by citizens' groups campaigning against the amendment, but the Yokkaichi city council, which had been entrusted with the signatures with the promise of submitting them to the city council, discarded them. On February 4, 1987, the governor of Mie Prefecture and the mayor of Yokkaichi City submitted to the government a written opinion in favor of the revision. It was after the submission of this opinion that the citizens' group learned that the signatures against the amendment had been discarded. The government judged that there would be no problem if Yokkaichi City agreed to the amendment and revised the law.[16]

In the case of air pollution, Japan has historically placed more emphasis on industry than on the health of its residents, and the health risks to residents have not been emphasized.

9.6 In conclusion

This chapter shows that the Japanese government has underestimated the damage caused by asthma, revised legislation to change its designation from "pollution asthma" to general "asthma," and exacerbated harmful outcomes by incorporating it into the universal health insurance system, making responsibility ambiguous.

Although we know that air pollution is detrimental to our health, quantifying its negative effects remains difficult. As this case shows, the difficulty of quantification makes it possible for culpable parties to avoid taking responsibility for the harm they cause. Regarding the PM2.5 problem, about which the science remains unclear, it is possible to consider what kind of attitude the country should take in response to potentially emerging health hazards.

If we focus on pollution patients, there are certainly people whose health as well as their employment and livelihoods are affected. Asthma patients are still on the rise. In the future, the number of air pollution patients will increase in every country in the world. This could happen in any country, as some people's lives are changed by the disease, there are restrictions on medical visits due to economic conditions (in some places, access to medical care is poor, to begin with), and medication compliance is not adhered to. Because of the variety of sources of air pollution, it is difficult to calculate the individual contribution, and the damage tends to be denied. In other words, from the perspective of the risk politics of air pollution in Japan, the damage is underestimated while the emphasis on the perpetrators by the state remains intact.

Anyone can develop asthma. We should not think that air pollution problems should be discussed by polluters or the government.

The air pollution problem should be considered as our problem, not as something that should be discussed by polluters or the government.

Notes

1 In the 2011 Fukushima Daiichi nuclear disaster, 28,000 people are still living outside Fukushima prefecture as of June 2021 due to concerns about health hazards caused by radiation exposure. The government has decided that it has been a long time since the accident occurred and that the possibility of radiation exposure in Fukushima Prefecture has decreased, and that it has voluntarily evacuated, so there is no support for housing expenses.
2 Environmental Restoration and Conservation Agency www.erca.go.jp/yobou/about/kougai.html
Retrieved February 11, 2021.
3 Japan has had a universal health insurance system since 1961, and currently it is generally 30% self-pay. If you are over 75 years old, a lower income will reduce your out-of-pocket rate.
4 Ministry of Health, Labor and Welfare 2017 Patient Survey Overview www.mhlw. go.jp/toukei/saikin/hw/kanja/17/index.html
Retrieved February 12, 2021.
5 The concentration of oil refineries on the Pacific side is due to the Taiheiyo Belt Zone concept, which prioritizes geographical conditions. This industrial zone, symbolized by the oil complex, was built on the Pacific side. In 1960, the Ikeda Cabinet presented an income doubling plan, including the Taiheiyo Belt Zone initiative. The Taiheiyo Belt Zone extends from South Kanto to Northern Kyushu.
6 From the 1979 Environmental White Paper of the Environment Agency.
7 In 2005, the Metropolitan Expressway Company Limited was privatized and changed to a joint-stock company.
8 This figure includes patients under the age of 18 whose medical expenses are unconditionally subsidized. The 2004 National Lifestyle Survey estimates that Tokyo is the home to about 197,000 bronchial asthma patients (Tokyo Metropolitan Air Pollution Medical Expenses Subsidy Review Committee, 2008).
9 The Ministry of Health, Labor, and Welfare explains livelihood protection as follows: "What is livelihood protection? For those who are still in need of living even if they utilize all their assets and abilities, we provide necessary protection according to the degree of need and guarantee a healthy and cultural minimum life. It is a system that compensates for independence." From the Ministry of Health, Labor and Welfare website. Https://www.mhlw.go.jp/stf/seisakunit-suite/bunya/hukushi_kaigo/seikatsuhogo/seikatuhogo/index.html
See September 30, 2020.

10 Recognizing the adverse effects of air pollution, the Tokyo Metropolitan Government enacted an ordinance in 1972 to independently subsidize the medical expenses of patients with four respiratory illnesses under the age of 18.
11 According to the 2010 Ministry of Health, Labor, and Welfare's Basic Survey on National Living, the average annual household income was 5,975,000 yen for those in their 40s and 7,359,000 yen for those in their 50s, for a total of 5,496,000 yen. In addition, the incomes of 61% were less than 5.49 million yen.
12 In Tokyo, information is disclosed once a week so that the media can report it. Whether or not the published information will be reported at the discretion of each media will be decided.
13 *Nihon Keizai Shinbun.* February 11, 2008.
14 *Nihon Keizai Shinbun.* February 11, 2008.
15 *Nippon Keidanren Times.* No. 2876, September 20, 2007.
16 Interview with Yoshiro Sawai, based on a survey of Yokkaichi pollution patients conducted between 2002 and 2004.

References

Hironao Ozaki, Manami Horihata, Masahumi Yokemoto, and Fu Zhe (2011) "Aratana taikiosen 'Mininteikanja' no kyuusaiseido sousetu ni muketa chousaken-kyuu houkokusho—Tokyotoiryouhijyoseijyourei no seisakukouka wo chuushinni" [Research Report for the Establishment of a New Relief System for Air Pollution 'Uncertified' Patients—Focusing on the Policy Effects of the Tokyo Metropolitan Medical Expenses Subsidy Ordinance], *Tokyokeizaidaigakugakujyutukenkyusenta-Working paper series 2010-E-02.*
Ken, Fujikawa (2018) "Hashigaki" [Foreword], in Ken Fujikawa & Masahumi Yokemoto (Eds.), *Hoshanou osen ha naze kurikaesarerunoka.* Tokyo, Toshindo, pp. i–vi.
Manami, Horihata (2009) "Mono wo iwanai kanjatachi—Yokkaichikougaikanja no genzai-" [Quiet Patients - The Present of Yokkaichi Pollution Patients-], in Oubirindaigakusangyoukenkyuusho *"Obirindaigaku sangyoukenkyuushohou"* 27, pp. 93–112.
Masahumi, Yokemoto and Manami Horihata (2003) "Mininteikanja no higaijittai ni kansuru jireihoukoku" [Case Report on the Actual Damage of Uncertified Patients],*"Kankyo to kougai" Iwanamishoten* 32(4), pp. 51–56.
Masashichi, Nishio (1979) "Nisankachisso ni kakaru hannteijyouken nado ni tuiteno senmon iinkai houkoku oyobi hugen ni kansuru koushuueisei gaku teki tachiba kara no gimon", [*Environmental Technology*] 8(7), pp. 708–710.
Nobuko, Iijima (1993) "Kankyoumonndai to higaishaundou" [Environmental Problems and Victim Movement], *Tokyo, Gakubunsha.*
Nobuko, Iijima, Watanabe, Shinichi, and Fujikawa, Ken (2007) "Kougaihigaihouchi no shakaigaku" [Sociology of Neglecting Pollution Damage], *Tokyo, Toshindo.*
Ryouichi, Yoshimura (2003) "Taikiosenkougaisoshou no nagare to Tokyososshou-hanketu" [Flow of Air Pollution Litigation and Tokyo Lawsuit], *"Kankyo to kougai" Iwanamishoten* 32(4), pp. 22–29.
Tokyototaikioseniryouhijyoseikentouiinnkai (July 2008) "Taikioseniryouhijyosei no seidokakudai ni kansuru houkokusho" [Report on Expansion of Air Pollution Medical Expenses Subsidy System]. https://www.fukushihoken.metro.tokyo.lg.jp/kankyo/kankyo_eisei/taiki/iryouhi/houkokusho.files/houkokusho.pdf.

Part IV

Contested risk constructions of air pollution

10 The individualization of responsibility for transboundary air pollution in Korea

Inkyoung Kim

Differing understandings of the causes of environmental problems have led to different views on how and who should address them (Conca et al., 1995), resulting in ongoing struggles within the international community over collective solutions to global environmental issues such as climate change and desertification. The framing of environmental deterioration as a byproduct of affluence at the United Nations Conference on the Human Environment in 1972 produced substantial concerns that the international environmental agenda could stifle economic development in the developing world. Conflicts over who should take the most or more responsibility to address environmental problems such as air pollution have played out in domestic as well as international environmental policy forums. Although all stakeholders, including governments, producers, consumers, and activists, need to take action to address environmental problems, who bears the most responsibility for solving the problem of air pollution remains a matter of ongoing dispute at the local, national, and international level.

Unlike land, the air as commons is not owned by any single individual or state: everyone shares it regardless of national borders. Thus, the responsibility for keeping air clean and breathable falls on various stakeholders, including national and local governments, corporations and industries, and individual households and communities. Governments can target or manage industries through regulatory measures, and the citizens who represent both the consumers and investors of the industries can legitimize the regulatory roles of the governments and urge corporations through their sustainable choices and activism to address their exposure to environmental and social risk (Siddiqui, 2018).

This chapter addresses air pollution governance in East Asia by investigating the contributions and limitations of the individualization of responsibility as a strategy for cosmopolitan risk governance. As it will show, the efforts of the Republic of Korea (hereinafter Korea) to reduce particulate matter, one of the biggest air pollution problems in East Asia, reveal a growing emphasis on individual consumption as a strategy for dealing with environmental pollution.

DOI: 10.4324/9781003211747-14

10.1 Individualization of responsibility

Responsibility is not a clearly defined concept. It can sit in a wide spectrum of definitions, from broad moral principles to legal obligations or duties (Kent, 2009). There are two recognized aspects of responsibility: matters of justice and law, implying duties and obligations (Bickerstaff & Walker, 2002; Caney, 2006; Singer, 2002, 2006), and a psychological phenomenon, implying moral values at the personal and societal levels (Auhagen & Bierhoff, 2000). For both aspects, responsibility is "necessarily socially mediated," as it involves duties for care and obligations based on relationships with others (Bickerstaff & Walker, 2002).

International law deals with state responsibility for internationally wrongful acts of a state. According to the International Law Commission, state responsibility arises with breach of an obligation, and taking responsibility for those acts requires enduring legal consequences flowing from that breach, such as cessation and assurances of non-repetition and reparation for injuries, including material and moral damage (United Nations, 2008). An iconic example regarding transboundary air pollution is the *Trail Smelter* arbitration between the USA and Canada in 1938 and 1941 (United Nations, 2006), which addressed what indemnity should be paid for damage caused by the Trail Smelter in Washington state, to what extent Trail Smelter should be required to refrain from causing further damage, and what measures or regime should be adopted or maintained. Those questions thus entail both "post-responsibility" for the effect or harm after the cause or activity and "ante-responsibility" for prospective and future effect or harm of the cause or activity (Birnbacher, 2000). In contrast to state or corporate responsibility, individualized responsibility for air pollution is related to ante-responsibility and places a duty of care on individuals instead of on states or corporations.

The individualization of responsibility can be defined as modern institution's imposing responsibility on individuals, resulting in changes in those individuals' attitudes as a sociological phenomenon (Beck, 2007). This concept, based on the neoliberal assumption that human well-being can best be advanced by maximizing entrepreneurial freedoms and limiting the role of the state within the market (Fahnbulleh, 2020), has become a "catchcry of politicians, bureaucrats and NGOs, including environmental organizations" and led governments around the globe to urge individuals to take greater responsibility for various social issues, including environmental pollution (Kent, 2009, p. 138).

Encouraging sustainable consumption has become a key strategy of states attempting to achieve sustainable development. According to the international Oslo Symposium in 1994, sustainable consumption refers to the "use of goods and services that respond to basic needs and bring a better quality of life, whilst minimising the use of natural resources, toxic materials and emissions of waste and pollutants over the life cycle" without compromising the needs of future generations (International Institute for Sustainable Development, n.d.).

Sustainable consumption efforts are supported by the concept of environmental citizenship, which defines individuals as social actors and calls upon them to understand and act with care for the environment (Hawthorne & Alabaster, 1999) by internalizing information about environmental issues, developing a sense of personal responsibility, and expressing that duty through consumption and community actions (Hobson, 2002). This concept thus demonstrates a move away from framing the citizen as a mere consumer, entailing "the emergence of an active citizen" who is mobilized by responsibility and duty "rather than the passive citizen, bounded by rights and privileges" (Hobson, 2002, p. 101).

Researchers have increasingly acknowledged the role of individuals as citizens and consumers in mitigating the effects of greenhouse gas emissions on climate change (Kent, 2009). Although the study of such behavior changes is now several decades old (Eden, 1993; Hinchliffe, 1996; Rüdig and Lowe, 1986), more practical approaches and grassroots initiatives have recently emerged, such as Earth Hour's annual campaign of switching off lights for an hour on the last Saturday of March, which has become "one of the world's largest grassroots movements for the environment" (Earth Hour, n.d.).

Some scholars, however, have been critical of efforts to individualize responsibility for their focus on consumption and lack of attention to structural influence by other agents, such as governments and businesses. Maniates (2001), for instance, dismisses the individualization of responsibility as "the notion that knotty issues of consumption, consumerism, power and responsibility can be resolved neatly and cleanly through enlightened, uncoordinated consumer choice" and warns that it underestimates "the nature and exercise of political power, or ways of collectively changing the distribution of power and influence in society" (p. 33).

Other studies of consumption behavior at the individual or household level have concluded that governmental and corporate efforts to promote public awareness of environmental risks are not adequate enough to have large environmental effects (Lorenzoni et al., 2007). As some have noted, people are citizens and active entities as well as mere consumers (Burgess et al., 2003; Malpass et al., 2007), and thus their consumption decisions are not merely economic but cultural, stemming from individual and societal norms and beliefs, making it difficult to "decipher how household exchanges and transactions are shaped within powerful networks of external agents, and by the cultural contexts from which they materialise" (Gibson et al., 2011, p. 5).

Furthermore, the individualization of responsibility can ignore other important causes of and solutions to environmental risks, including corporate social responsibility. This critique is particularly relevant to East Asian countries. The prevalent Asian culture of governance is based on developmentalism and authoritarian environmentalism (Ahlers and Shen, 2018) and requires a "policy regime with a command and control mechanism" (Siddiqui, 2018, pp. 233–234). For East Asian countries that suffer severe transboundary air pollution, therefore, the individualization of responsibility can pose great challenges for solving the problem effectively.

The rest of this chapter examines the degree to which individuals have been asked to take responsibility for tackling worsening air pollution in Korea, investigating both legislative and administrative measures and addressing the effectiveness of those efforts. Because the problem of particulate matter is exacerbated by both local and foreign factors, the following sections will first discuss how the Korean government has addressed local factors domestically and then how it has addressed transboundary sources of particulate matter internationally.

10.2 Air pollution in Korea

Particulate matter has become the most serious air pollution issue in East Asia, including PM10 (particles with an aerodynamic diameter of 10 microns or smaller, about one-seventh the width of a human hair) and PM2.5 (finer particles with aerodynamic diameter of 2.5 microns and smaller, about one-twenty-fifth or -thirtieth of a human hair) (US EPA, 1995). According to WHO, particulate matter is a "common proxy indicator for air pollution" and "affects more people than any other pollutant" (WHO, 2018). WHO warns that exposure to PM2.5 increases health risks, including "cardiovascular and respiratory disease, and cancers," and causes 4.2 million premature deaths worldwide every year (WHO, 2018). Since 2000, Korea has reported the highest share of mean population exposed to PM2.5 concentration among OECD countries (OECD, 2020; Trnka, 2020).

Despite these alarming levels, Korea has improved its annual average concentration levels of various air pollutants. According to the Ministry of Environment of Korea (hereinafter MOEK), Seoul has gradually improved PM10 concentrations since 2002 and stabilized PM2.5 concentrations since 2008 (MOEK, 2015, p. 47). But while the annual average concentration levels for PM2.5 decreased in all regions in 2015–2018, this decline has not been notable. The annual average from December to May actually rose during this period, as did the number of warnings and alerts, from 72 in 2015 to 177 in 2017 (OECD, 2020). Indeed, the frequency and duration of high concentration episodes increased during this period (Table 10.1).

Table 10.1 Number of continuing hours of unhealthy and very unhealthy levels of PM2.5, 2015–2018

	2015	2016	2017	2018
Nationwide	16.2	24.4	17.7	26.5
Seoul	12.1	13.6	18.1	20.5
Kyoung-gi region	17.7	16.9	23.1	25.1

Source: *NCCA (2020b, p. 7).*

10.3 Individualizing responsibility for particulate matter in Korea

As one of the three main actors along with governments and industries, citizens can play a critical role in addressing environmental pollution. This section investigates the extent to which the Korean government has used legislative and administrative efforts to individualize responsibility for transboundary air pollution, particularly in response to increasing concerns regarding the particulate matter.

10.3.1 Legislation

Five major legislative acts have been particularly relevant to Korea's efforts to address air pollution.[1] As reported in Table 10.2, this study categorized the 239 articles of those five acts according to the responsibilities for pollution that they assigned to governments, businesses, and citizens.

The Framework Act on Environmental Policy (enacted 1990 and last amended 2019) is intended to preserve the environment and prevent environmental damage by defining the rights and duties of citizens and enterprises and the obligations of state and local governments (Articles 1 & 2). It outlines the obligations of state and local governments in Article 4, of business entities in Article 5, and the rights and duties of citizens in Article 6 (National Law Information Center, 2017a). Article 7 articulates the principle that industries or individuals who cause pollution or damage by their business or activities are liable for expenses incurred in restoring the damage and preventing expected pollution. Although this act broadly defines the responsibilities of different entities, most of the articles elaborate upon the specific roles of state and local governments rather than the roles of industries and citizens. For example, they stipulate that national and local governments shall create environmental standards (Articles 12 & 13), establish national environmental preservation plans (Articles 14–21), implement environmental impact assessment (Article 41), devise policies on dispute mediation and damage relief (Articles 42–44), establish a special account for environmental improvement accounting (Articles 45–53) and financial measures (Article 55–57), and establish a Central Environmental Policy Committee and an Environmental Preservation Association (Articles 58–59). In fact, of the 59 articles, 52 (88.1%) address the actions to be taken by the national and local governments; only two (3.4%, Articles 5 & 30) directly address the responsibilities of business entities, and only one (1.7%, Article 6) the duties and rights of citizens.

The Clean Air Conservation Act (enacted 1995 and last amended 2019) sets up very detailed responsibilities (National Law Information Center of Korea, 2019a). Of its 76 articles, 18 (23.7%) assign national and local governments the task of monitoring and assessing air pollutants (Articles 3 & 4), predicting and announcing air pollution levels (Article 7) by operating the National Center for Integrated Control of Air Quality (newly inserted in 2013 Article 7.3), issuing air pollution alerts to the relevant area (Article 8), and providing financial and technical support for manufacturers of low-emission motor vehicles, exhaust gas reduction devices, and low-emission engines (Article 47). In this act, however, the largest number of articles—43 (56.6%)—addresses the role of businesses, such as by determining permissible emission levels (Article 16), requiring the installation of measuring devices (Article 32), imposing emission charges, and managing fossil fuel use (Articles 41 & 42), fugitive dust levels (Article 43), and volatile organic compounds (Articles 44 & 45). The act declares that vehicle manufacturers shall comply with permissible levels for pollutants (Article 46) and average emission quantities (Article 50.2). Only eight of the articles (10.5%) address the responsibility of individuals, more specifically that of owners of motor vehicles to ensure compliance with the permissible emission levels (Article 57), the permitted life span and emission levels of air pollutants of their vehicles (Article 58), restrictions on engine idling (Article 59), and inspections of exhaust gases (Article 63).

Of the 34 articles of the Special Act on the Improvement of Air Quality in Seoul Metropolitan Area (enacted 2003), 16 (47.1%) address the responsibilities of state and local governments, 12 (35.3%) those of businesses, and 4 (11.8%) those of citizens (National Law Information Center of Korea, 2017b). Article 2 defines "metropolitan area" as encompassing Seoul Special Metropolitan City, Incheon Metropolitan City, and the Gyeonggi-do region and is home to almost half the South Korean population. The articles regarding the government stipulate that the Minister of Environment shall formulate and implement a master plan to reduce seven air pollutants, including PM10 and PM2.5 (Article 8), and promote the supply of low-pollution motor vehicles (Articles 23–24) and that the Mayor of Seoul may subsidize costs to scrap specific vehicles that exceed permissible emission levels (Article 27). Those regarding businesses require all businesses to comply with the total volume control of pollutants (Articles 14–22), car dealers to obtain government approval of their annual supply plans for low-pollution motor vehicles in accordance with national supply standards (Article 23.3), and all exhaust gas reduction devices or low-pollution engines to be certified (Article 26). The responsibilities of citizens focus heavily on vehicle owners, including requiring owners of specific diesel vehicles to install exhaust gas reduction devices or replace the engine with a low-pollution engine to comply with permissible emission levels (Article 25).

This act was replaced with the Special Act on the Improvement of Air Control in Air Control Zones (enacted 2019), which also expands the regulation for air pollution beyond the Seoul Metropolitan area to include any area with serious air pollution (Article 2) (National Law Information Center of Korea, 2021). This act further expands the responsibilities of business and citizens; of its 39 articles, 15 (38.5%) address the former and 8 (20.5%) the latter, whereas those of government were reduced to 11 (28.2%%). The increasing number of articles addressing businesses and citizens include more specific regulations for reducing vehicle emissions in areas with serious air pollution issues, such as requiring diesel vehicle owners to comply with more strengthened emission standards

(Article 26.1) and prohibiting preschool buses and transportation trucks from using diesel vehicles (Article 28).

The Special Act on the Reduction and Management of Fine Dust (enacted 2018 and last amended 2020) targets particulate matter by reducing fine dust emissions and substances (Article 1) (National Law Information Center of Korea, 2019b). Most of the 31 articles in this act address the specific responsibilities of state and local governments: 25 (80.6%) obligate governments to formulate comprehensive plans for particulate matter management (Article 7); establish a Special Policy Committee (Articles 10 & 11), Office for Clean Air (Article 12), and National Air Emission Inventory and Research Center (Article 17); enhance international cooperation (Article 14); "ascertain the route,

Table 10.2 Articles assigning responsibilities to different actors in five major acts

Act	Date Enacted	Date Last Amended	Total No. of Articles	Government	Business	Citizens
Framework Act on Environmental Policy	January 8, 1990	November 26, 2019	59	52 (88.1%)	2 (31.4%)	1 (1.7%)
Clean Air Conservation Act	December 29, 1995	November 26, 2019	76	18 (23.7%)	43 (56.6%)	8 (10.5%)
Special Act on the Improvement of Air Quality in Seoul Metropolitan Area	December 31, 2003	April 2, 2019 Abolished	34	16 (47.1%)	12 (35.3%)	4 (11.8%)
Special Act on the Reduction and Management of Fine Dust	October 14, 2018	March 31, 2020	31	25 (80.6%)	3 (9.7%)	2 (6.5%)
Special Act on the Improvement of Atmospheric Environment in Atmospheric Management Area	April 2, 2019	—	39	11 (28.2%)	15 (38.5%)	8 (20.5%)
Total			239	122 (51%)	75 (31.4)	23 (9.6%)

Note: Some articles are double-counted, as they include several paragraphs that address different roles of different actors.

concentration, etc. of long-range transported fine dust" (Article 15); and take emergency reduction measures during high concentration levels (Article 18.1). Only 9.7%—a general description (Article 4) and only one article (Article 18)—address compliance by business, such as changing or adjusting their operating hours and rates (National Law Information Center of Korea, 2019b). The responsibilities of citizens are broadly defined in Article 5, which states that they "shall endeavor to reduce and manage fine dust, etc. that are generated in their daily lives and cooperate in the policies implemented by the State and local governments for the reduction and management of fine dust," and in Article 18.3.3, which stipulates that a "person in receipt of a request for emergency reduction measures . . . shall comply therewith unless there is a compelling reason not to do so."

As this close investigation of the five major acts reveals, state and local governments are assigned the biggest number of responsibilities for reducing air pollution, and more specifically particulate matter, in Korea. It demonstrates a range of legislative responses to socio-environmental problems, and that the "ante-responsibility" of individuals is limited largely to their vehicle ownership rather than energy use within their households.

10.3.2 *Administrative measures*

Korea also has developed a series of administrative measures to tackle particulate matter issues, which include tightening air quality standards and enhancing enforcement mechanisms by increasing local inspections and strengthening national supervision. In 2014, the Korean government introduced the Fine Dust Forecasting and Alarm System to provide the public with real-time warnings of high concentrations of particulate matter.[2] Moon Jae-In's administration (2017–2022) has prioritized the fight against particulate matter and adopted a package of particulate matter countermeasures that include the Comprehensive Plan on Fine Dust Management in September of 2017 and the Fine Dust Reinforcement Measures in January of 2018 (Ministry of Culture, Sports, and Tourism of Korea, 2020). The Korean government officially designated the country's problems with particulate matter a "social disaster" on February 15, 2019, when the Special Act on the Reduction and Management of Fine Dust was enacted (NCCA, 2020a).

In accordance with Article 18 of the Special Act on the Reduction and Management of Fine Dust, emergency reduction measures are enforced to reduce particulate matter levels when concentration levels at the municipal or provincial level are higher than the guidelines. For example, in the Seoul and Kyoung-gi metropolitan areas, emergency mitigation measures implemented on November 7, 2018, were estimated to have reduced PM2.5 emissions by 6.8 tons, or about 4.7% (MOEK, 2018a; NCCA, 2020a).

The goals of the 2017 Comprehensive Plan on Fine Dust Management are to reduce local emissions by 30% by 2020 and the number of "bad" air quality days by about 66% by 2022. The Korean government identified goals or at least strategies for four major sectors of emission: power generation, industries, transportation, and everyday surroundings (Joo, 2018). The power generation sector is directed to reduce emissions by 25% by 2022 through decreasing the share of coal-fired power generation. Industry must reduce its emissions by 43% by 2022 through various measures including intensive inspection and monitoring. The transportation sector needs to phase out old diesel vehicles and is encouraged to use environment-friendly cars. For everyday surroundings, the Korean government controls blind spots such as construction sites and illegal incinerations, distributes road-cleaning vehicles, and extends urban forests.

To deal with domestic sources of particulate matter, the Korean government has imposed primary responsibility for reducing particulate matter upon fossil fuel-fired power plants and industries. According to a recent study, the industrial sector is the greatest domestic contributor to annual concentrations of surface PM2.5 in Korea, accounting for about one-fifth of annual average concentrations, followed by the residential, commercial, and power generation sectors (Yim et al., 2019). In the metropolitan areas, construction equipment and power plants accounted for 22% and 11% of the PM2.5 concentration, respectively, whereas nationally, industrial facilities accounted for 41%, construction equipment for 17%, and power plants for 14% in 2016 (MOEK, 2016).

Due to the high contribution of businesses, construction sites, and power plants nationwide, the government has adopted strengthened management policies for the December–May period, during which the country's 42 coal- and heavy oil-fired power plants are allowed to operate at only 80% of their usual level for that time of year (MOEK, 2018c). In March 2019, the Ministry of Trade, Industry, and Energy announced that they would extend that regulation to a total of 60 power plants, permanently shut down six old power plants over the next three years, and partially suspend the operation of 48 coal power plants for 7–45 days during the high concentration season (*Yonhap News Agency*, 2019a).

Because the government is aware of the limitations of their efforts to cut particulate matter emissions through managing state-run power firms, it has called for more cooperation from the four industries—steel, petrochemical, oil refining, and cement—most responsible for high concentrations of particulate matter (*Yonhap News Agency*, 2019a). The role of industries in air pollution, however, has not been widely discussed by the public even though Korea's leading industries are energy-intensive and account for more than 61.4% of total energy consumption, more than transportation (18.9%), residential (9.4%), commercial (7.5%), and public sectors (2.8%) combined (Hong et al., 2019). According to the NCCA, because these mitigation measures are "mandatory" only for public sectors and only in metropolitan areas, they do not affect the biggest sources of PM2.5: industrial facilities and construction equipment (NCCA, 2020a, p. 9).

Individuals are assigned responsibility for reducing emissions to tackle particulate matter issues in Korea mostly through vehicle ownership, particularly through ownership of diesel vehicles, which account for 29% of PM2.5 concentrations in metropolitan areas and 11% nationwide (MOEK, 2016). To urge individuals to take responsibility for particulate matter, the Korean government has offered both sticks and carrots.

For sticks, it has restricted the operation of high-emission vehicles based on a car rating system announced in May 2018, which grades all cars in terms of their emissions (Lim, 2018). This new system offers economic incentives, including parking discounts and toll reductions, to owners of vehicles given the highest grades of 1 and 2 and places certain limitations on the operation of vehicles receiving the lowest grades of 4 and 5. In 2018, there were about 269 million grade 5 vehicles in Korea, about 99% of them diesel-fueled (MOEK, 2018b), which were not allowed to operate on roads during periods when particulate matter concentrations are expected to be high (MOEK, 2018c). Since March 2020, all grade 5 emission vehicles have been barred from roads in the metropolitan areas (*Yonhap News Agency*, 2019b).

Beginning December 1, 2019, the Seoul Metropolitan Government imposed regulations on grade 5 vehicles, particularly in the Green Transport Zone (inside the Seoul City Wall) from 6 am to 9 pm (Seoul Metropolitan Government, 2019). This includes a fine of 250,000 won (about US$200) per violation per day under Article 48 of the Enforcement Decree of Sustainable Transportation Logistics Development Act. The government aims to phase out diesel vehicles in the public sector by 2030 and reduce diesel vehicles in the private sector by encouraging a transition to LPG-run vehicles (MOEK, 2018c).

For carrots, since 2005, the government has provided subsidies to individuals willing to take such emission reduction measures as installing diesel particulate filters and retiring old vehicles early. In 2018 alone, 16,845 vehicles installed emission reduction devices, 220 vehicles installed redesigned engines, and 11,411 vehicles were retired early, which together were estimated to have reduced PM2.5 by 2,085 tons, about 6.2% of the 33,698 tons emitted annually by diesel-fueled vehicles (MOEK, 2018a). In 2019, the subsidy rate increased from 50% to 60% of value and the targeted number of early retirements doubled to 300,000 units, with the goal of scrapping 77% of the 2.86 million diesel vehicles built prior to 2006 by 2022 (Argus, 2019).

The Korean government has also provided incentives for individuals to purchase eco-friendly vehicles. In 2020, the ministries of finance and environment changed a state subsidy scheme to allow larger subsidies for buyers of energy-efficient cars (*Korea Herald*, 2020), including one-time subsidies for the purchase of an electric vehicle of up to 18.2 million won (about US$15,700) and up to 42.5 million won (about US$36,700) for a hydrogen-powered vehicle (*Korea Herald*, 2020). These measures seemed aligned with the Implementation Plan for the 3020 Renewable Energy, which aims to increase the portion of renewable energy from the current level of less than 3% to 20% of power generation by 2030 (Hong et al., 2019).

Other administrative measures taken to reduce air pollution include the establishment of several new governmental agencies. In accordance with the Special Act on the Reduction and Management of Fine Dust, the National

Council on Climate and Air Quality (NCCA) was launched on April 29, 2019, to identify and evaluate agendas on fine particulate air pollution (NCCA, 2020a). In August of that year, the National Information Center for Fine Dust was also established under the auspices of the Ministry of Environment (MOEK, 2019). The Korean government also created the prime minister-led Special Policy Committee on Fine Dust comprising relevant ministers and civilian experts to take charge of policies on fine dust reduction (Ock, 2019). In February 2019, that committee presented two sets of behavioral guidelines or protocols to be followed during periods of high levels of particulate matter, one for various groups of individuals and the other for the general public. The first of these includes specific response instructions for toddlers, students, the elderly, patients with pre-existing conditions, outdoor workers, and agrarians highlighting what institutions and individuals need to do during days with high levels of particulate matter, such as preparing for outdoor classes at school at the institution level and wearing facial masks and washing hands frequently at the individual level (Special Policy Committee on Fine Dust, n.d.). The second sets out a Promise of the Ten Public Actions, including riding a bike, using public transportation, purchasing eco-friendly motor vehicles, and not idling engines. Some of these recommendations have also been supported by such official actions as the Seoul metropolitan government's shutting down 424 public parking lots and issuing fines of 100,000 won (about US$83) to 4,500 vehicles for idling engines (*MBC News*, 2019).

10.4 Assessing the individualization of responsibility

The analysis of the five major acts finds that they have not done much to embed individual responsibility in the law, as only 9.6% of a total of 239 articles have addressed the role of individuals in air pollution mitigation and management, in comparison to more than half of the total number of articles addressing the role of governments and almost a third addressing that of businesses. Yet it also finds that the various administrative measures adopted in support of these acts, including both sticks and carrots for vehicle owners, have led to greater individualization of responsibility as a sociological phenomenon.

This psychological or sociological individualization of responsibility can be seen in the results of a recent survey by the Hyundai Research Institute (Min, 2019), in which the respondents identified emissions by diesel vehicles (10.5%), coal-power plants and energy generation-related combustion (6%), and industry (3.2%) as the main domestic sources of particulate matter in Korea. This public perception that vehicles contribute the most and industries the least to the total amount of particulate matter is aligned with (if considerably lower than) their contribution to total emissions in Korea's metropolitan areas but very different from those shares at the national level, as discussed earlier in this chapter. On the national level, therefore, this public perception of individuals' responsibility for concentrations of particulate matter through their transportation choices exaggerates their actual contribution.

10.5   Transboundary air pollution in East Asia

Even though the Korean government has emphasized individual responsibility through legislative and administrative measures to address local sources for air pollution, it is fully aware of the limitations of individualization of responsibility due to significant sources of transboundary pollution and public demands for more international strategies beyond domestic measures. In the survey by the Hyundai Research Institute (Min, 2019), 78.3% of the respondents stated that most particulate matter in Korea's air originates from neighboring countries, especially China. The respondents indicated that most of the responsibility for solving this problem lies with the government. Asked to identify strategies for improving Korea's air quality, 67.5% prioritized international cooperation and research between the Korean and Chinese governments, 10.3% strengthening domestic standards, and 9.3% managing the demand for diesel and other automobiles.

In fact, numerous studies have investigated the serious impacts of the influx of air pollutants emitted from large, fast-growing cities and industrial complexes in China and other foreign sources on Japan (e.g., Aikawa et al., 2010; Kaneyasu et al., 2014), Korea (e.g., Han et al., 2008; Heo et al., 2009; Kim et al., 2017; Vellingiri et al., 2016), and East Asia as a region (e.g., Hou et al., 2019; Wang et al., 2019). Regarding transboundary air pollution in Korea, a joint study by Korea's National Institute of Environmental Research (NIER) and the US National Aeronautics and Space Administration, called KORUS-AQ, assessed that 48% of PM2.5 in Korea was caused by foreign sources (34% from China, 9% from North Korea, and 5% from Japan and elsewhere) from May to June 2016 (MOEK, 2017). According to a 2019 study, up to 70% of the annual average PM2.5 concentrations in Japan and South Korea came from foreign countries, particularly China (Yim et al., 2019). According to the Joint Research Project for Long-range Transboundary Air Pollutants in Northeast Asia (LTP) that includes China, Japan, and Korea, China's contributions account for about 32.1% of PM2.5 concentrations in major Korean cities (Secretariat of Working Group for LTP, 2019).

In addition to the legislative efforts to embed individual responsibility for better air quality and the various administrative measures to implement the law, the Korean government has developed both legislative and administrative measures to improve air quality through international cooperation, particularly with China. In 2015, the Clean Air Conservation Act was amended to move from issue-specific measurements for yellow dust to more comprehensive measurements for transboundary air pollution. Article 2.22 introduced the term "long-range transboundary air pollutants" to the act for the first time, defining them as "air pollutants prescribed by Ordinance of the Ministry of Environment, which have transboundary impacts on multiple countries through long-range movements after the creation of yellow dust, dust, etc." (National Law Information Center of Korea, 2019a). Articles 13 and 14 replaced "yellow-dust" with "long-range transboundary air pollutants" to recognize pollution issues beyond yellow dust. Article 15 significantly expanded the specificity of international cooperation, replacing the previous brief statement that "the Government shall make endeavors to cooperate with relevant nations to prevent damage caused by yellow dust and other air pollutants effecting countries" with commitments to seven specific measures:

1   Holding, supporting, and participating in various events, such as international conferences and academic conferences;
2   Participating in exchanges of technology and human resources with relevant countries and international organizations;
3   Supporting research and disseminating findings on transboundary air pollutants;
4   Conducting education and public relations campaigns on transboundary air pollutants within the international community;
5   Raising financial resources to prevent damage caused by transboundary air pollutants;
6   Establishing an air pollution monitoring system in Northeast Asia;
7   Taking any other matters necessary for international cooperation.

The Korean government also has been involved in much more specific and rigorous environmental cooperation with China regarding particulate matter. During Chinese President Xi Jinping's state visit to Korea in July 2014, the two governments issued a joint statement ensuring joint monitoring of air quality and sharing data and technology on dust collection, denitrification, and desulfurization (MOEK, 2014). In 2015, they agreed to share real-time data on PM2.5 to help Korea enhance the accuracy of its Fine Dust Forecasting and Alarm System. Weekly forecasts were piloted in the second half of 2019 and have since allowed preliminary reduction measures, such as road cleaning and bidiurnal vehicle rotation in public sectors, to be conducted the day before a high concentration of fine particles is predicted (MOEK, 2018c). In 2017, the Korea-China Environmental Cooperation Plan 2018–2022 led to the creation in 2018 of a Korea-China Environmental Cooperation Center in Beijing and the Korea-China Cooperation Initiative for the Demonstration of Environmental Technologies for Fine Dust Reduction to advance cooperation on technology transfer between businesses in Korea and China.

On November 26, 2019, Korea also amended the Framework Act on Environmental Policy to call for national and local governments' active participation in international cooperation to address transboundary air pollution (Article 27) and to identify the International Environmental Cooperation Center (IECC) as the specific organization to lead international environmental cooperation (Article 27.2) (National Law Information Center of Korea, 2019c). In 2020, the Korea Environmental Industry and Technology Institute was named as the host of the IECC Center, which is charged with enhancing international environmental cooperation and providing leadership for global environmental efforts through policy analysis, data collection and analysis, and collaborative research projects regarding international environmental agreements and regulations (Sun, 2020).

Korea's international cooperation with countries in East Asia to tackle transboundary air pollution issues started in the 1990s (Kim, 2007; Kim, 2014). However, it was not until when the Korean government amended the Clean Air Conservation Act to address various environmental issues of transboundary air pollution beyond yellow dust in 2015 that it introduced the

> the term "long-range transboundary air pollutants" to its legal system. After international efforts to understand transboundary air pollution through several non-legally binding agreements for information sharing and joint research for more than two decades, the Korean government has also been able to identify significant contribution of domestic sources to deteriorating air quality. At a press conference in Korea in 2015, Greenpeace also called for more governmental efforts to address domestic sources of its particle-laden smog. Greenpeace proposed to reduce Korea's dependence on coal plants for energy production and to develop more renewable energy sources.
>
> (Jung, 2015)

To manage domestic sources, the Korean government may have contributed to public perception of exaggerated individuals' contribution to and responsibility for particulate matter by developing domestic policies for reducing the number of smoggy days, such as increasing the price of diesel and regulating restaurants in ways to decrease sulfur and nitrogen emissions in Korea. Governmental efforts to ask the public to do their part in tackling particulate matter have also confused the public by proposing zig-zag policies on diesel. In contrast to recent efforts to discourage the purchase and operation of diesel vehicles, in 2015 the Korean government actually offered incentives for diesel vehicles, leading to an 8% increase in diesel consumption that year (Jaganathan and Jang, 2016). The promotion of diesel technology as "clean diesel" was common internationally until the Volkswagen emissions scandal, known as "dieselgate," revealed that those figures were inflated (Parloff, 2018).

10.6 Conclusion

As strategies for cosmopolitan risk governance to tackle air pollution, the Korean government has promoted various strategies at the extremes of the spectrum, individual responsibility and international cooperation. In addition to developing international cooperation on transboundary air pollution and collaboration with industries to reduce domestic sources of particulate matter, the Korean government has promoted individual responsibility for reducing particulate matter emission among its populace. This individualized responsibility has reaped some reduction of particulate matter emissions in Korea, particularly through emergency measures such as restricting the operation of old diesel vehicles and encouraging environmentally friendly behaviors among citizens on a daily basis.

Korea has a "centralised system of environmental governance" albeit with multilevel governance by subnational, such as provincial and local, governments (Trnka, 2020, p. 15). For the MOEK, as the environmental authority to establish and implement its own environmental policies and legislative development since 1994 (MOEK, n.d.), over 60% of its budget is spent on support to local governments to improve local inspection and enforcement capacity of environmental regulations (Trnka, 2020). Perhaps as a result of an institutional culture that assigns primary responsibility for environmental policies regarding air pollution to MOEK, the role of the private sector in reducing particulate matter remains unclear and nascent at most. Most measures proposed for industry have been recommendations without teeth, whereas the public sector, particularly state-owned power plants, and individuals operating diesel vehicles have been tightly monitored and regulated. To successfully tackle particulate matter issues, the Korean government will need to pay much more attention to industry, which is the country's highest energy user and thereby biggest emitter of air pollutants.

Because individuals are community members who are breathing common air, government policies should also expand efforts to individualize responsibility for the quality of that air beyond their current focus on the transport sector. Individuals should be encouraged to view themselves as environmental citizens with a sense of responsibility for air pollution not simply as consumers but also as voters, investors, and activists empowered to drive governmental policy changes and eco-friendly industrial practices. By using stakeholder engagement to promote a more active role for businesses, making greener consumer choices, and encouraging greater international cooperation in the region, Korean citizens can play a vital role in efforts to deal with transboundary air pollution in East Asia.

Notes

1 Although the Framework Act on Low Carbon, Green Growth (enacted 2010 and last amended 2019) also identifies such responsibilities, it is not included in this analysis because its attention to issues of climate change, energy, greenhouse gases, and traffic systems has only broader relevance to the specific issue of air pollution addressed in this chapter (National Law Information Center of Korea, 2018a). Several other acts also regulate air quality but focus on smaller and more specific issues, such as the Indoor Air Quality Control Act (enacted 1996 and last amended 2019), which addresses air pollution issues in public-use facilities and mass transit vehicles (National Law Information Center of Korea, 2018b), and the School Health Act (enacted 1995 and last amended 2020), which addresses ventilation and air quality monitoring devices in schools (National Law Information Center of Korea, 2020).

2 At the city level, Seoul had already implemented the Fine Dust Forecasting and Alarm System for PM10 since 2005 and for PM2.5 since 2013 and integrated its system into the national standards since 2016 (Seoul Metropolitan Government, n.d.).

References

Ahlers, A. L., & Shen, Y. (2018). Breathe easy? Local nuances of authoritarian environmentalism in China's battle against air pollution. *China Quarterly*, 234, 299–319.

Aikawa, M., Ohara, T., Hiraki, T., Oishi, O., Tsuji, A., Yamagami, M., Murano, K., & Mukai, H. (2010). Significant geographic gradients in particulate sulfate over Japan determined from multiple-site measurements and a chemical transport model: Impacts of transboundary pollution from the Asian continent. *Atmospheric Environment*, 44, 381–391.

Argus. (2019). *South Korea to expand EV subsidy allocation.* www.argusmedia.com/en/news/1968432-south-korea-to-expand-ev-subsidy-allocation

Auhagen, A. E., & Bierhoff, H.-W. (2000). Responsibility as a fundamental human phenomenon. In A. E. Auhagen & H.-W. Bierhoff (Eds.), *Responsibility: The many faces of a social phenomenon* (pp. 1–8). Routledge.

Beck, U. (2007). Beyond class and nation: Reframing social inequalities in a globalizing world. *British Journal of Sociology*, 58(4), 679–705.

Bickerstaff, K., & Walker, G. (2002). Risk, responsibility, and blame: An analysis of vocabularies of motive in air-pollution(ing) discourses. *Environment and Planning A*, 34, 2175–2192.

Birnbacher, D. (2000). Philosophical foundations of responsibility. In A. E. Auhagen & H.-W. Bierhoff (Eds.), *Responsibility: The many faces of a social phenomenon* (pp. 9–22). Routledge.

Burgess, J., Bedford, T., Hobson, G., Davies, G., & Harrison, C. M. (2003). (Un)sustainable consumption. In F. Berkhout, M. Leach, & I. Scoones (Eds.), *Negotiating environmental change: New perspectives from social science* (pp. 261–292). Edward Elgar.

Caney, S. (2006). Cosmopolitan justice, rights and global climate change. *Canadian Journal of Law and Jurisprudence*, 19(2), 255–278.

Conca, K., Alberty, M., & Dabelko, G. D. (Eds.). (1995). *Green planet blues: Environmental politics from Stockholm to Rio.* Westview Press.

Earth Hour. (n.d.). *Our mission.* www.earthhour.org/our-mission

Eden, S. E. (1993). Individual environmental responsibility and its Role in public environmentalism. *Environment and Planning A*, 25(12), 1743–1758.

Fahnbulleh, M. (2020). The neoliberal collapse: Markets are not the answer. *Foreign Affairs*, 99(1), 38–43.

Gibson, C., Head, L., Gill, N., & Waitt, G. (2011). Climate change and household dynamics: Beyond consumption, unbounding sustainability. *Transactions of the Institute of British Geographers*, 36(1), 3–8.

Han, Y.-J., Kim, T.-S., & Kim, H. (2008). Ionic constituents and source analysis of PM2.5 in three Korean cities. *Atmospheric Environment*, 42, 4735–4746.

Hawthorne, M., & Alabaster, T. (1999). Citizen 2000: Development of a model of environmental citizenship. *Global Environmental Change*, 9(1), 25–43.

Heo, J., Hopke, P. K., & Yi, S.-M. (2009). Source apportionment of PM2.5 in Seoul, Korea. *Atmospheric Chemistry and Physics*, 9, 4957–4971.

Hinchliffe, S. (1996). Helping the earth begins at home: The social construction of socio-environmental responsibilities. *Global Environmental Change*, 6(1), 53–62.

Hobson, K. (2002). Competing discourses of sustainable consumption: Does the 'rationalisation of lifestyles' make sense? *Environmental Politics*, 11(2), 95–120.

Hong, J.-H., Kim, J., Son, W., Shin, H., Kim, N., Lee, W. K., & Kim, J. (2019). Long-term energy strategy scenarios for South Korea: Transition to a sustainable energy system. *Energy Policy*, 127, 425–437.

Hou, X., Chan, C. K., Dong, G. H., & Yim, S. H. L. (2019). Impacts of transboundary air pollution and local emissions on PM2.5 pollution in the Pearl River Delta region of China and the public health, and the policy implications. *Environmental Research Letters*, 14, 034005.

International Institute for Sustainable Development. (n.d.). *Symposium: Sustainable consumption, 1994*. http://enb.iisd.org/consume/oslo004.html#top

Jaganathan, J., & Jang, R. (2016, February 18). Rising diesel demand in South Korea puts brakes on exports. *Reuters*. www.reuters.com/article/us-southkorea-diesel-demand/rising-diesel-demand-in-south-korea-puts-brakes-on-exports-idUSKCN0VR0NB

Joo, H.-S. (2018). Comprehensive plan on fine dust Management. *Korea Environmental Policy Bulletin*, 40(2). https://library.kei.re.kr:444/dmme/img/001/015/005/KEPB_46%ED%98%B8_%EB%AF%B8%EC%84%B8%EB%A8%BC%EC%A7%80.pdf

Jung, M. (2015, March 4). Greenpeace spares China from blame for fine dust. *The Korea Times*. www.koreatimes.co.kr/www/news/nation/2015/03/116_174608.html

Kaneyasu, N., Yamamoto, S., Sato, K., Takami, A., Hayashi, M., Hara, K., Kawamoto, K., Okuda, T., & Katakeyama, S. (2014). Impact of long-range transport of aerosols on the $PM_{2.5}$ composition at a major metropolitan area in the Northern Kyushu area of Japan. *Atmospheric Environment*, 97, 416–425.

Kent, J. (2009). Individualized responsibility and climate change: 'If climate protection becomes everyone's responsibility, does it end up being no-one's?'. *Cosmopolitan Civil Societies Journal*, 1(3), 132–149.

Kim, B.-U., Kim, C. B., Kim, H. C., Kim, E., & Kim, S. (2017). Spatially and chemically resolved source apportionment analysis: Case study of high particulate matter event. *Atmospheric Environment*, 162, 55–70.

Kim, I. (2007). Environmental cooperation of Northeast Asia: Transboundary air pollution. *International Relations of the Asia-Pacific*, 7(3), 439–462.

———. (2014). Message from a middle power: Participation by the Republic of Korea in regional environmental cooperation on transboundary air pollution issues. *International Environmental Agreements: Politics, Law and Economics*, 14(2), 147–162.

Korea Herald. (2020, January 20). S. Korea changes subside scheme for eco-friendly autos. www.koreaherald.com/view.php?ud=20200120000674

Lim, E. (2018). *Ministry of Environment announced a new car rating system*. www.climatescorecard.org/2018/05/ministry-of-environment-announced-a-new-car-rating-system/

Lorenzoni, I., Nicholson-Cole, S., & Whitmarsh, L. (2007). Barriers perceived to engaging with climate change among the UK public and their policy implications. *Global Environmental Change*, 17, 445–459.

Malpass, A., Barnett, C., Clarke, N., & Cloke, P. (2007) Problematizing choice: Responsible consumers, sceptical citizens. In M. Bevir & F. Trentmann (Eds.), *Governance, consumers and citizens: Agency and resistance in contemporary politics* (pp. 231–256). Palgrave Macmillan.

Maniates, M. F. (2001). Individualization: Plant a tree, buy a bike, save the world? *Global Environmental Politics*, 1(3), 31–52.

MBC News. (2019). Trapped in particulate matter, unbearable without a mask. (In Korean) http://imnews.imbc.com/replay/2019/nwdesk/article/5637989_24634. html?xtr_cate=LK&xtr_ref=r11&xtr_kw=N&xtr_area=k22&xtr_cp=c4

Min, J. (2019). Analysis of Korean's perception on particulate matter. (In Korean) *Weekly Economic Review.* Hyundai Research Institute. www.hri.co.kr/board/reportView.asp?firstDepth=1&secondDepth=1%20&numIdx=30120

Ministry of Culture, Sports and Tourism of Korea. (2020). *Find dust comprehensive measures.* www.korea.kr/special/policyCurationView.do?newsId=148864591&pWise=sub&pWiseSub=B12

MOEK (2014). *Press release: Full implementation of Korea-China environmental cooperative projects for the era of environmental partners* (In Korean). http://me.go.kr/home/web/board/read.do;jsessionid=1rfTtch67KHQWlkk1IlgzNzU.mehome1?pagerOffset=5680&maxPageItems=10&maxIndexPages=10&searchKey=title&searchValue=&menuId=286&orgCd=&condition.fromDate=2014-07-03&boardId=355028&boardMasterId=1&boardCategoryId=&decorator=

———. (2015). *ECOREA: Environmental review 2015.* Korea.

———. (2016). *Air quality measures: South Korea.* www.unescap.org/sites/default/files/Session%201.1.%20Youngsook%20Yoo_ROK.pdf

———. (2017). *Press release: Result of KORUS-AQ.* www.me.go.kr/home/web/board/read.do?menuId=286&boardId=803050&boardMasterId=1

———. (2018a). *Press release: Reduction of about 2,000 tons of particulate matter through emission reduction measures for vehicles on roads* (In Korean). http://me.go.kr/home/web/board/read.do;jsessionid=EVb4AZlgE7iIhU33+evj-sqJ.mehome1?pagerOffset=1940&maxPageItems=10&maxIndexPages=10&searchKey=&searchValue=&menuId=286&orgCd=&boardId=927770&boardMasterId=1&boardCategoryId=&decorator=

———. (2018b). *Starting promotion of grade 5 of emission gas from December 1.* (In Korean). http://me.go.kr/home/web/board/read.do;jsessionid=FK8XBKHdZROTVDQ7ij+lC-eE.mehome1?pagerOffset=1960&maxPageItems=10&maxIndexPages=10&searchKey=&searchValue=&menuId=286&orgCd=&boardId=926180&boardMasterId=1&boardCategoryId=&decorator=

———. (2018c). *Press release: Inclusive environmental policy that everyone enjoys a clean environment.* http://eng.me.go.kr/eng/web/board/read.do;jsessionid=tfi4FnwdfgsCLlvbk1eADYB0.mehome1?pagerOffset=0&maxPageItems=10&maxIndexPages=10&searchKey=titleOrContent&searchValue=2014&menuId=461&orgCd=&boardId=931710&boardMasterId=522&boardCategoryId=&decorator=&firstItemIndex=

———. (2019). *Press release: Opening the national information center for fine dust, support for fine dust policies* (In Korean). https://me.go.kr/home/web/board/read.do?pagerOffset=0&maxPageItems=10&maxIndexPages=10&searchKey=&searchValue=&menuId=286&orgCd=&boardId=1144520&boardMasterId=1&boardCategoryId=39&decorator=

———. (n.d.). *History of ME.* https://eng.me.go.kr/eng/web/index.do?menuId=471

National Law Information Center of Korea. (2017a). *Framework Act on environmental policy.* Act No 14532. http://law.go.kr/lsInfoP.do?lsiSeq=191130&chrClsCd=010203&urlMode=engLsInfoR&viewCls=engLsInfoR#0000

———. (2017b). *Special Act on the improvement of air quality in Seoul metropolitan area.* Act No. 15274. http://law.go.kr/lsInfoP.do?lsiSeq=199873&chrClsCd=010203&urlMode=engLsInfoR&viewCls=engLsInfoR#0000

———. (2018a). *Framework act on low carbon, green growth*. Act No. 16133. http://law.go.kr/lsInfoP.do?lsiSeq=206348&chrClsCd=010203&urlMode=engLsInfoR&viewCls=engLsInfoR#0000

———. (2018b). *Indoor air quality control act*. Act No. 15583. http://law.go.kr/lsInfoP.do?lsiSeq=203193&chrClsCd=010203&urlMode=engLsInfoR&viewCls=engLsInfoR#0000

———. (2019a). *Clean air conservation act*. Act N. 16266. http://law.go.kr/lsInfoP.do?lsiSeq=206702&chrClsCd=010203&urlMode=engLsInfoR&viewCls=engLsInfoR#0000

———. (2019b). *Special act on the reduction and management of fine dust*. Act No. 16303. http://law.go.kr/lsInfoP.do?lsiSeq=208137&chrClsCd=010203&urlMode=engLsInfoR&viewCls=engLsInfoR#0000

———. (2019c). *Framework act on environmental policy* (In Korean). http://law.go.kr/lsInfoP.do?lsiSeq=211571&efYd=20200527&ancYnChk=0#0000

———. (2021). *Special act on the improvement of air control in air control zones*. Act No. 17983. https://law.go.kr/LSW/lsInfoP.do?lsiSeq=230779&chrClsCd=010203&urlMode=engLsInfoR&viewCls=engLsInfoR#0000

———. (2020). *School health act*. Act 17472 (In Korean). http://law.go.kr/lsInfoP.do?lsiSeq=220859&efYd=20200912&ancYnChk=0#0000

NCCA. (2020a). *Report on the current state and solutions for fine particulate air pollution: Abridgement*. www.ncca.go.kr/cmn/board/BBSMSTR_000000000005/1044bbsDetail.do

———. (2020b). *Report on the current state and solutions for fine particulate air pollution* (In Korean). www.ncca.go.kr/cmn/board/BBSMSTR_000000000005/1044bbsDetail.do

Ock, H.-J. (2019, February 14). What is S. Korea doing to combat fine dust pollution? *The Korea Herald*. www.koreaherald.com/view.php?ud=20190214000358

OECD. (2020). *Air quality and health: Exposure to PM2.5 fine particles -countries and regions, OECD environment statistics (database)*. https://doi.org/10.1787/96171c76-en

Parloff, R. (2018, February 6). How VW Paid $25 billion for 'dieselgate'—and got off easy. *Fortune*. https://fortune.com/2018/02/06/volkswagen-vw-emissions-scandal-penalties/

Rüdig, W., & Lowe, P. (1986). The withered 'Greening' of British politics: A study of the ecology party. *Political Studies*, 34, 262–284.

Secretariat of Working Group for Long-range Transboundary Air Pollutants in Northeast Asia. (2019). *Summary report of the 4th Stage (2013–2017) LTP Project*. https://www.me.go.kr/home/file/readDownloadFile.do?fileId=184686&fileSeq=1

Seoul Metropolitan Government. 2019. *Seoul to control g5 vehicles of emission gas in green transport zones from December 1*. http://english.seoul.go.kr/seoul-to-control-grade-5-vehicles-of-emission-gas-in-green-transport-zones-from-dec-1/

Seoul Metropolitan Government. n.d. *Common sense of fine dust*. https://bluesky.seoul.go.kr/finedust/common-sense/page/10?article=745

Siddiqui, A. I. (2018). Can responsible corporate behavior clean up the haze? In E. Quah & T. S. Tan (Eds.), *Pollution across borders: Transboundary fire, smoke and haze in Southeast Asia* (pp. 225–235). World Scientific Publishing.

Singer, P. (2002). *One world: The ethics of globalization*. Text Publishing Company.

———. (2006). Ethics and climate change: A commentary on MacCracken, Toman and Gardiner. *Environmental Values*, 15, 415–422. Special Policy Committee on

Fine Dust. (n.d.) *Promise of the ten public actions.* www.cleanair.go.kr/dust/dust/dust-action02.do

Sun, B.-K. (2020, September 24). Establishment of international Environmental Cooperation Center. *Kukto-Ilbo.* www.ikld.kr/news/articleView.html?idxno=223997

Trnka, D. (2020). Policies, regulatory framework and enforcement for air quality management: The case of Korea. *OECD environmental working paper* No. 158. www.oecd-ilibrary.org/docserver/8f92651b-en.pdf?expires=1601935014&id=id&accname=guest&checksum=D48E26BDE637862C16E642966609E4FB

United Nations. (2006). *Reports of international arbitral awards: Trail Smelter case* (United States, Canada). https://legal.un.org/riaa/cases/vol_III/1905-1982.pdf

———. (2008). *Draft articles on responsibility of states for internationally wrongful acts, with commentaries.* https://legal.un.org/ilc/texts/instruments/english/commentaries/9_6_2001.pdf

US EPA. (1995). *AIRTrends 1995 Summary.* https://www3.epa.gov/airtrends/aqtrnd95/pm10.html

Vellingiri, K., Kim, K.-H., Lim, J.-M., Lee, J.-H., Ma, C.-J., Jeon, B.-H., Sohn, J-R., Kumar, P., & Kang, C.-H. (2016). Identification of nitrogen dioxide and ozone source regions for an urban area in Korea using back trajectory analysis. *Atmospheric Research,* 176–177, 212–221.

Wang, M. Y., Yim, S. H. L., Wong, D. C., & Ho, K. F. (2019). Source contributions of surface ozone in China using an adjoint sensitivity analysis. *Science of the Total Environment,* 662, 385–392.

WHO. (2018). *Ambient (outdoor) air pollution.* www.who.int/news-room/factsheets/detail/ambient-(outdoor)-air-quality-and-health

Yim, S. H. L., Gu, Y., Shapiro, M. A., & Stephens, B. (2019). Air quality and acid deposition impacts of local emissions and transboundary air pollution in Japan and South Korea. *Atmospheric Chemistry and Physics,* 19, 13309–13323.

Yonhap News Agency. (2019a, March 6). *Gov't to limit operation of coal power plants to curb fine dust pollution.* https://en.yna.co.kr/view/AEN20190306007400320

———. (2019b, November 1). *(LEAD) Gov't unveils tough anti-dust measures, as fine dust blankets much of Korea for 2nd day.* https://en.yna.co.kr/view/AEN20191101002951315

11 Science in air pollution politics
School allocation and regulatory control disputes on SNCC, Taiwan

Wen-Ling Tu

11.1 Introduction

As the environmental issues have become increasingly more critical, the Government is in a dire need of new policy knowledge and tools to handle the complex human and environmental relationships. Typical environmental policymaking process involves external experts and scientific instruments to help clarify the questions and answers to make a good judgment. The modern environmental decision-making processes are embedded with the various scientific models and experiments as well as expert meetings to reach final conclusions. However, focusing on a linear scientific model for policymaking has a great limitation. Taking the air pollution governance as an example, the academic research-based epidemiological surveys, air pollution monitoring empirical data, as well as the local people's life experiences altogether seem able to patch the relationship between industrial pollution and health damage. But many local life experiences and knowledge may not be acknowledged by the governmental agencies when they need to make a specific administrative discretion. How should we understand the role of science in air pollution politics, especially when there are great scientific disputes in identifying pollution risks?

In this chapter, I will use the case study of the pollution disputes of the Sixth Naphtha Cracker Complex (SNCC), the largest petrochemical complex in Taiwan, to demonstrate the science disputes in pollution politics. By organizing and reviewing the data collected from the news clippings, public documents, and the participation observation in the relevant meetings, I will particularly focus on the school relocation controversies of the Syucuo Branch of Ciaotou Elementary School (SBCES) at Mailiao Township, Yunlin County.

11.2 Political nature of science in risk policymaking

Contemporary science emphasizes on the acquisition and systematization of positive knowledge. Science often involves truth-telling and objectivity, so scientific knowledge is mostly perceived as reliable and persuasive in political settings (Jasanoff, 2005). It is the core basis of policymaking and the pillars of administrative legitimacy and authority. The administrations have become even

DOI: 10.4324/9781003211747-15

more dependent on the knowledge and information provided by the scientific experts to make clear policy guidance for the changing environment. As Jasanoff (1990) noticed, the scientific advisory board experts played an indispensable role and were deemed as "the fifth branch" of the government since the 1970s.

This reliance, application, and reverence of science echo Lasswell's observation in *The Policy Orientation* that policy science must be "problem oriented, multidisciplinary, methodologically sophisticated, theoretically sophisticated, and value oriented" (Smith and Larimer, 2016). The approach emphasizes the importance of pursuing scientific rationality for policymaking. Heavy reliance on scientific analytical tools such as risk assessment and cost-benefit analysis has essentially turned these tools into guidelines for policymakers. To maintain policymaking objectivity and impartiality, Fisher (2007: 11–13) argued that science needs to be kept separate from politics, policy, and the law so as to ensure its integrity.

The expectation that science can provide the correct knowledge and clarify complex problems is often based on the prerequisite that the science experts are value-neutral, and scientific advice can provide the "objective" knowledge for policymaking. But different from pure scientific research where verifications and hypothesis proofs do not need to concern with any policymaking time pressures, the science-related public decision-making is often forced to be made while a vast amount of information still remained unknown or the scientific evidence has not yet reached a consensus. The judgment criteria for policy science are often volatile, controversial, and political.

In light of the potential risks that have become increasingly difficult to control and exceeded the understanding of contemporary scientific knowledge, the risk management that primarily relies on science is subject to significant limits. Some studies have indicated that scientific tools can only master very little amount of facts in the natural society due to the high degree of changes and uncertainties (Hinchliffe, 2001). Sometimes the number of questions posed by science is more than the number of answers (Sarewitz, 2000). The science, technology, and society (STS) studies are particularly concerned about the interwoven relationships between the scientific technology knowledge production processes and the social factors. This approach argues that the knowledge gained based on the universal principle without *contextualized* practices can only present part of the facts. How science is positioned in the social power structure may affect research results and scientific knowledge is often not as rational and objective as it claimed (Davis, 2002).

As Douglas (2005: 156) noticed, scientists cannot avoid value judgment in their work. But "consideration of ethical values in places of needed scientific judgment pose no threat to objectivity as such." In her point of view, the problem of objectivity arises when "values take the place of evidence" or "lead one to simply ignore evidence that runs contrary to a desire outcome." The idea of keeping "scientific objectivity" may suppress a clear discussion of legitimate value judgments on the selection of methodologies, data quality, and interpretations that further violate the standard of democracy. To hold scientists accountable, she suggested scientists be "more explicit in their work concerning where judgments are made and how they made them."

The STS studies have noted that the overall direction of scientific research agendas tended to conform to the wills of *capital interests*. Hess (2007) observed that task-oriented academic program grants have become more focused on the key industries for national and regional economic development under the call of technology transfer and industrial innovation. While environmental sustainability-related studies are critical, they produced limited results. He developed the "undone science" concept and discussed the subject of systematic knowledge nonproduction by the science academia community. Frickel et al. (2010: 445) further define "undone science" as "areas of research identified by social movements or civil society organizations as having potentially broad social benefit, but are left unfunded, incomplete, or generally ignored." The concept helps us to more keenly observe the role of the institutional factors in the shaping of scientific research, and understand why some knowledge cannot be systemically produced in the institutional matrix of state, industry, and social movements. How science is embedded in the social network deserves further examination.

Some studies have further pointed out how scientific knowledge production may be influenced by social-economic powers. In particular, those who oppose the regulations often use the "manufacturing uncertainty" and "junk science" strategies to delay or eliminate regulatory actions, which stress the impossibility of using science to confirm the causality relationships between the industrial hazards and the health risks (Michaels and Monforton, 2005; Michaels, 2008). Oreskes and Conway (2010) found that the scientists in the conservative camp have used their past reputations to manufacture doubts of science for the climate change issue to delay positive policy actions. Some renowned scientists, such as the celebrated British epidemiologist Professor Sir Richard Doll, was found holding a paid consultancy with chemical firms. His reviews often underestimated the role of chemicals in causing cancers and were largely used by the manufacturers' trade association.[1] The case demonstrated that the financial sponsorship and political agenda behind the scientific controversies need to be well-reviewed.

The money flow driven by the industrial/commercial interests influences not only the sponsorship of scientific projects but also community risk perceptions and interpretations, which often mingle a complicated political and economic calculation of local benefits. Ottinger (2013) studied some fence line communities near the refineries in the USA and found that residents have contradictory points of view on the pollution issues and how the problems should be depicted publicly. On one hand, "painting the problems in the community too starkly, or too publicly, created the risks of driving away potential homebuyers," (p. 95) company investment, or potential job opportunities. On the other hand, some community members challenge the industrial scientists' and engineers' data assertion that there was nothing wrong in the living environment (Ottinger, 2010).

Relying on science to perform risk assessments poses significant limits, but such limits are often overlooked in the environmental regulatory policies or standards (Davis, 2002). The aforementioned studies have cast questions that scientific knowledge has become more closely aligned with the economic and political powers. The dynamics of scientific knowledge creation and application in the policy decision-making process are subjects for further examination. In the following section, I will use the case study of SBCES relocation controversies to situate the science in local politics. Recognizing the financial sponsorship and political-economic power in play in the local context, I will highlight the power of capital and localism in Taiwanese/Asian pollution politics.

11.3   Measuring air pollution, health risk, and SBCES relocation

The SBCES was built at Mailiao Township, Yunlin County of Taiwan in 2013, which is only 0.9 km away from the SNCC. It was moved to this location from its old address because the original campus was small and dilapidated, and the advent of the SNCC had increased the number of students and made campus expansion necessary. Formosa Plastics Group, the main shareholder of the complex which had an annual turnover accounting for 9.6% of Taiwan GDP (TWD 1.7 trillion in 2018), had originally promised to build feedback constructions such as local hospitals, schools, elderly care centers, etc., before the construction of the SNCC in 1994; but this promise was not fulfilled until 2010 because Formosa Plastics owned no land to build these "promised" facilities. Here is a brief history of how Formosa Plastics obtained the land for these facilities. Formosa Plastics was supposed to construct a 500-meter wide isolation waterway as part of the industrial isolation zone for the SNCC but made a waterway area reduction proposal (from 500 meters to 100 meters) to the Industrial Development Bureau (IDB) of the Ministry of Economic Affairs (MEA). The proposal was rejected, but the IDB of MEA approved a compromised plan for a 200-meter waterway alongside a 300-meter wide green-belt.

A few years later, Formosa Plastics turned to the Yunlin County Government (YCG) for permission to build the aforementioned "promised" facilities on the state-owned windbreak forest lands within the said 300-meter isolation green-belt, and the YCG approved the request. As a result, institutions such as the Chang Gung Memorial Hospital, the SBCES, the Taichung Investigation Bureau, and the Taichung Customs Services are now located inside the 500-meter area originally designated as the industrial isolation zone of the SNCC. The construction costs for the SBCES were up to NT$100 million, and Formosa Plastics sponsored over NT$70 million. The construction of the SBCES started in 2010, and the SBCES was opened in 2013.

The health risk issues had become a matter of concern since the SNCC construction started, so Formosa Plastics promised to conduct the health risk assessments for specific hazardous air pollutants and submit the assessment reports to the EPA and Environmental Protection Bureau (EPB) of YCG for future reference. However, the first relevant report was not submitted until as late as 2008. Moreover, the IDB has commissioned the NCKU Research Development Foundation to conduct an "Environmental and Health Exposure and Risk Assessment Study in Yunlin Offshore Industrial Zone" study in 2005, and the results indicated that the health risks tended to be high for the people living in the vicinity of the SNCC. However, this study report was treated as a "restricted data" for internal reference only. It was not until 2009 this report was first disclosed by the media. In response to the information concealment, the IDB issued a press release explaining that there are many carcinogenic factors in the environment, and further epidemiological data is required to determine the causal relationship between air pollution and cancer (Tu and Shih, 2014).

In 2009, Professor C.C. Chen accepted a YCG commissioned an investigation and found that the VOCs discharged from the SNCC were significantly related to the local people's health. From 2010 to 2011, fire incidents frequently occurred in the SNCC and triggered the community's concerns toward the SNCC health risk issues. After the SBCES was activated at Mailiao in 2013, the National Health Research Institute (NHRI) and Prof. C.C. Chan at National Taiwan University (NTU) jointly conducted a two-year "Epidemiological Study on the Schoolchildren Living near the Sixth Naphtha Cracker Complex (SNCC)—the Preliminary Survey on the Vicinity of Vinyl Chloride (VCM) Exposure and Health Condition of School-aged Children in Mailiao Township" (hereafter "the VCM exposure study"). The preliminary results of the VCM exposure study published in 2014 showed that the urinary thiodiglycolic acid (TdGA) levels of the SBCES students (the closest school to SNCC) were significantly higher (at 193.06 µg/g-creatinine) than those at other schools. TdGA is a major metabolite of VCM, and high levels of TdGA were found in the human urine samples of students from the neighboring schools within a 9km radius. The study also found that the TdGA levels were 4 times higher during school time than those during the summer vacation. The results made the VCM factory in the SNCC a suspicious target of major pollution source.

On August 22, 2014, the YCG, MHW, MOE, and EPA have jointly announced to temporarily relocate the schoolchildren of the SBCES to the Ciaotou Elementary School (CES). However, the term "temporary placement" seemed to signify that the government has not yet approved the NHRI research results. They used the "emergency avoidance" measure to implement the school relocation and never conducted any further risk communications. These rushed actions have buried the seeds of even more intensified controversies for the future.

One year after the school relocation, the parents started to complain about having to leave the new SBCES and be transferred back to the CES. The old school is crowded and has a poor learning environment. The parents and the local politicians thus oppose the YCG to maintain the school relocation decision and believe that if the SNCC caused the pollution, the entire Mailiao Township would have been polluted. So why is it that only the SBCES students are relocated? The true solution would be to provide investigation reports and order the SNCC to relocate. On July 29, 2015, the parents of the SBCES students announced that they would move their kids back to the SBCES during September regardless of the YCG's decision, so the YCG acquiesced and allowed the students to move back to the SBCES. The Risk Society and Policy Research Center (RSPRC) of NTU issued a statement stating that the precautionary principle should be adopted for the school return decision based on the health and safety considerations of the schoolchildren. The RSPRC further urged the government to actively rectify the SNCC air pollution problems in order to save the children from air pollution exposure and resolve the root of the problem.[2]

The schoolchildren stayed in the SBCES for the entire 2015 academic year. On August 16, 2016, Prof. Chan spoke in the social group forum held by the Executive Yuan and once again explained that the schoolchildren of SBCES were exposed to a polluted environment. He suggested that Premier should issue an order to relocate the school. The Executive Yuan then consulted with the relevant agencies to learn about the NHRI research progress in order to verify the impacts of environmental pollution on the health of the schoolchildren as soon as possible. On August 22, 2016, the MHW convened an expert review meeting to verify the following items based on the NHRI research results: (1) The closer the school is to the SNCC, the higher the TdGA levels compared to those of the students from other schools. When the schoolchildren of SBCES were relocated to CES, their TdGA levels were immediately lowered to the same levels as those of the CES students. (2) The sources of TdGA are diverse. The research has made adjustments for the hepatitis B, vitamin B consumption, passive cigarette smoke exposure, and home to source distance factors; but cannot completely rule out the VCM/ethylene dichloride, etc. (3) The annual monitoring data from the EPA shows that the mean VCM level was 2.2 ppb with a maximum level of 165 at SBCES. Although this level has not exceeded the national standard, it was still higher when compared to that of the USA. (4) A follow-up study is needed. However, preventive relocation is recommended to protect the health of the schoolchildren. Subsequently, the government decided to relocate SBCES to Fong Long elementary school (FLES).[3]

The YCG Magistrate Lee then convened a contingency meeting to reach the following resolutions: (1) The central government should invite experts to discuss whether there is a direct causal relationship between the pollution source VCM and its biomarker TdGA as well as the human health. (2) Regarding the significantly high level of TdGA found in students in SBCES, the central and local governments have recommended to temporarily relocate the schoolchildren from SBCES to FLES in order to keep the children away from the high-risk exposure.[4]

This led to strong protests from the parents, who believe that the decision was devoid of any prior communication and did not respect the local people. They did not trust the NHRI research results and requested to have another research unit conduct the investigation. Approximately 300 people protested in front of the Executive Yuan on August 24, 2016. The protesters then negotiated with the relevant agencies and reached an agreement on September 1, 2016, to commission the National Cheng Kung University Hospital (that the parents trusted) to collect urine samples from the SBCES schoolchildren for analysis and conduct three more subsequent tests. With a "third-party" performing the tests, the parents then agreed to relocate their children to the CES.[5] However, the environmental groups questioned the integrity of NCKU Professor C. C. Lee who took over the tests because he was commissioned by Formosa Plastics to implement the SNCC health risk assessment plan. They argued that the professor should recuse himself due to a conflict of interest.[6]

On February 7, 2017, the Health Promotion Administration (HPA) convened an "SBCES environmental monitoring and urine investigation expert meeting" to discuss the findings of the study. The HPA emphasized that the effort was not a "research project," so the data would only be given to the individuals who participated in the tests. However, they would "explain the VCM monitoring results and the TdGA level in the schoolchildren's urine samples, and invite experts to review the data."[7] After the expert meeting, the HPA issued a press release and indicated that the TdGA concentration levels in the schoolchildren of SBCES are still high, but the TdGA concentration levels in the schoolchildren's urine samples were caused by numerous factors. A comparison of the urine test results with the EPA's environmental monitoring data found no significant correlation between the TdGA concentration levels and the VCM exposure. The cause of the high TdGA concentration levels must further be investigated. The HPA requested the environmental protection authorities to continue to monitor the VCM levels in the ambient air. The meeting did not determine whether the schoolchildren should be relocated back to the SBCES in the future. The undetermined conclusion led the schoolchildren of SBCES back to the new campus in August 2017 after three times of school relocation controversies.

11.4 Dilemma of applying science to air pollution governance and health policy decisions

11.4.1 Incomplete science, manufacturing doubts, and political fights

The SBCES relocation controversy exemplifies the contradictory actions taken by the government in handling the industrial pollution and health risk problems. As mentioned above, the results of Prof. Chan's research indicated that the operations of the SNCC do indeed degrade the air quality of the neighboring areas. The urine samples of those who lived within 10 km of the SNCC for over 5 years showed elevated metal concentration levels (the positive indicators of petrochemical industrial toxins), and their lung, liver, and kidney functions as well as cardiovascular systems have also been affected (Chan et al., 2011). The news report revealed the research result not only won the Society of Publishers in Asia Award honor but also aroused society's concerns for the SNCC pollution.[8] With the support of the scientific results, the Yunlin County Taixi Township residents living near the SNCC formed the "Yunlin County Taixi Township SNCC Pollution Damage Joint Claim Self-help Group" and filed a class-action lawsuit against the SNCC in 2015 to claim health damage liability compensations.

The joined research conducted by NTU and NHRI (2014) seemed to further clarify that there is a positive correlation between the health risk of residents and their residential location distance to the SNCC. As the NHRI issued a public official statement, jointly signed by five prestigious scientists, to clarify the characteristics of VCM, the significance of the TdGA levels detected in the urine samples of schoolchildren, and the possible sources. The primary evidences proposed were: Comparison of urinary TdGA level for participating 343 schoolchildren from 5 elementary schools in the vicinity of the SNCC, the median urinary TdGA levels for students attending the SBCES was the highest (193.06 µg/g-creatinine), significantly higher than FAES (101.05 µg/g-creatinine), CES (121.32 µg/g-creatinine) and MES (113.52 µg/g-creatinine). The research also adjusted for passive cigarette smoke exposure, vitamin B consumption, hepatitis infection status, BMI, their father's employment history, and home to VCM/PVC source distance. The statement indicated that the children studying in the neighborhood of approximately 9 km away from SNCC may have been exposed to VCMs. While the SBCES is only 0.9 km away, the schoolchildren there possibly experiencing higher VCM exposures. Their research findings were published in the 2016 Environmental Research Journal (Huang et al., 2016).

However, the report also mentioned some research limitations as follows: (1) They did not measure the ambient air levels of VCM or Ethylene dichloride (EDC) in any subject during their sampling period. Therefore, they cannot provide direct evidence regarding the sources of TdGA. (2) They did not collect all the urine samples on the same day. (3) The sampling period for 5 school comparisons was during autumn and spring when the wind direction was similar. But the repeated sampling for TdGA in SBCES was autumn and summer to compare differences in TdGA exposure between school time and vacation. (4) They did not utilize a detailed food questionnaire to evaluate the possible effects of food products or containers. (5) They could not rule out potential exposure from drinking water (Huang et al., 2016: 572).

Formosa Plastics took full advantage of the aforementioned research limitations to challenge its results. It argued that the NHRI international publication contains numerous uncertainties and contradictions, such as "TdGA came from diverse sources and does not represent VCM;" "the Association Advancing Occupational and Environmental Health (ACGIH) literature information showed that the correlation between TdGA and VCM cannot be verified when the VCM concentration level in the environment is below 5ppm;" and "the international literatures also indicated that the Bis (2-chloroethyl) ether (BCEE), a solvent used to make pesticides or other cleaners, can also produce TdGA in the metabolic processes of animals." In addition, the level of TdGA influenced by passive cigarette smoke exposure, automobile gas exhaust, and Vitamin B consumption may all exceed that caused by VCM, so the VCM level in the environment must also be tested as a reference value while the urine samples are being taken." The company further pointed out that the VCM value detected by Formosa Plastics is extremely low (0.24 ppb, far below the perimeter standard value of 220 ppb), and the liver function index (Alanine Aminotransferase/ ALT) of the schoolchildren showed no significant difference.[9] These doubts were primarily used to challenge the direct link between SNCC's VCM plant and the TdGA found in the schoolchildren's urine samples. The contested arguments presented that the determination of causality still remained inconclusive and more scientific research would be required to clarify the findings.

These doubts were also supported by some public health experts, who believed that "since the source of TdGA is not solely VCM, how can we prove that the TdGA metabolites in the bodies of schoolchildren came from SNCC pollution leaks? This doubt is not scientifically unreasonable."[10] Prof. C.C. Lee at NCKU (who had accepted the SNCC commissions to conduct long-term studies) and NHRI researcher Dr. C.P. Wen (with American petrochemical company working experiences) had both performed VCM monitoring inside the SNCC's VCM factory, and they unsurprisingly dismissed the correlation between the VCM and the TdGA. Wen stated that the EPA monitoring data did not find any VCM exposure value, so the question of whether the TdGA was produced by the SNCC's VCM plant should not be over-interpreted. He further added that there is no "research information indicating the TdGA found in the urine samples of schoolchildren may cause cancer in the future."[11] Prof. Lee argued that using the TdGA found in the urine samples to evaluate the effects of VCM should only be applicable to the workers with higher exposure levels, and not to schoolchildren who are also affected by numerous different environmental background factors. To understand the correlations between VCM and the TdGA found in the urine samples of schoolchildren, we must compare the data with that of the SNCC's VCM monitored by the EPA.[12]

In June 2017, Prof. C.C. Lee published the urine test results commissioned by the HPA, which indicated that the TdGA concentration in schoolchildren is still extremely high. However, because the result of the VCM test conducted by the EPA on school campus and homes one day before the urine sample collection was non-detected (ND), it therefore inferred that the TdGA found in the SBCES teachers and students came from other sources. The questionnaires filled out by the parents also indicated that after excluding passive cigarette smoke exposure, hepatitis B, and other food variables; the more eggs the students ate, the higher the TdGA levels in their bodies. However, because the questionnaire was filled out by the parents and the process did not undergo the Institutional Review Board (IRB) review process, "the exam was not regarded as a study, just a general health check." Lee's test results indicated that the high TdGA levels may be caused by eating eggs and has nothing to do with the VCM, which generated even greater controversies. In addition to the environmental groups' belief that the research was wrong and has no referential value,[13] the media also questioned why the TdGA background value investigation for the average children of Taiwan is only one-third compared to that of the schoolchildren in Mailiao Township. Is the problem only attributable to the eggs eaten by the SBCES students?[14]

Regardless of whether the funding sponsorships or the academia-industry collaborative relations may create research bias, we can see several scientific uncertainty controversies as follows: 1) Several studies have found higher than general concentration levels of TdGA in the urine samples of schoolchildren, but the scholars have different opinions as to where the source came from or whether TdGA in the urine samples can be the indicator of cancer risk. 2) It seems difficult to prove the VCM exposure in the vicinity of the VCM plant. There is no systematic, continuous, and accurate VCM measurement. 3) There is no conclusive agreement among the scientists about the safety standard of VCM exposure, in particular for high-sensitive groups such as schoolchildren. These controversies indicated that further studies are required, including expanded research designs, integrated environmental testing data, and repeated sampling comparison based on the previous research, in order to clarify the specific pollution sources and their health impacts. However, the government did not take a more comprehensive approach to scientific investigations to face the controversies and challenges. The scientific disputes that manufacture doubts in combination with the government inaction rendered no further understanding but fierce political fights.

11.4.2   Inconsistent policy actions weakened scientific claims and furious parents searched the third party for "impartial" science

As mentioned, the government announced the SBCES relocation three times within three years, and in most cases, the announcements were made less than a week before schools started. The resettlement processes were rushed and the parents have been dissatisfied and angry. Although the relocation decisions appeared scientifically rational and conformed to the preventive and precautionary principles, they led to even bigger storms of controversies. Formosa Plastics held a press conference to counterattack the policy, pointing out that its plants have fully complied with the national environmental standards.[15] The parents held up white banners and issued a press release to urge the government to cancel the order. The parents made four demands:

> Allow the students to stay and study in the SBCES; regulate and control the possible pollution sources via strengthened supervision instead of school relocation; hold public hearings to listen to the public concerns; and absolute refusal to be relocated to the other campus.

The parents also accused the NHRI study of

> not transparent, inconsistent sampling periods, the first phase contained urine and blood tests while the second phase contained only urine tests, and unable to confirm the source of the pollution from the SNCC. They requested that a third-party be commissioned to continue with the urine

sample tests and demanded the EPA to measure the ambient air levels of the VCM in order to triangulate the findings. They would agree to the relocation plan only if a correlation can be established.[16]

While the parents were strongly questioning the validity of the NHRI research results, they reached a negotiation agreement with the YCG to require a "third-party to recollect the samples and re-conduct the tests" in order to confirm the TdGA levels. Prof. C.C. Lee at NCKU was retained to assist with the school-children's TdGA tests. The first urine sample collection was completed on September 2 and 3 of 2016, and monitoring was conducted once a month for the next three months thereafter.

The political influence of the scientists' voices is rather alarming. The government has then promised more detections, studies, and environmental monitoring; which apparently did not help to resolve the disputes. Making the school relocation decision within a short period of one to two weeks has caught the school-children, the parents, and the schools by surprise. The administrative policymaking appeared no consideration about the consistency in setting the schoolchildren protection measures, strengthened petrochemical plant leakage control, expansion of the research scope based on the data already known, conduction of community risk communication, encouraging the schools to develop air pollution monitoring science education, etc. The school relocation program that everyone focused upon lacks infrastructural support. For most parents, the compromised resettlement location is also within the high-risk district identified by the experts. They questioned that even with the school relocation, two-third of the students' lives are still near the SNCC. So the school relocation is not the solution to the problem. Under the same logic, the whole village must be relocated in order to remedy the problem.[17]

Subsequently, the MHW arranged to have Prof. Lee and The NHRI collect urine samples from schoolchildren simultaneously in response to the local protests. Thus, the seemingly short-term conflict resolution has turned into more land mines of future controversies. Because the original risk prevention decision was based on the NHRI investigation research, the government's acquiesce to "redo the tests" signified an indirect recognition that the "NHRI cannot be trusted" and has weakened the scientific claims that the original decision was based on. However, the bigger questions of the re-test are why only the urine samples of the SBCES schoolchildren are being retested and what is the purpose for conducting the study? The HPA indicated that, "This plan is not a study, so it does not need to be 'academically' reviewed. It is only done in response to the parent's expectations of a fair investigation."[18] Nevertheless, the unclear purpose for the tests fraught with conflict of interest contentions can only further intensify the politics of science and start another round of disputes after the results have emerged.

As mentioned, the environmental groups distrusted Prof. C.C. Lee because he had accepted long-term funding from Formosa Plastics. Prof. Lee stressed that his risk assessment research has prompted Formosa Plastics to reduce a large amount of pollution. While he "did research for Formosa Plastics, he has never advocated for Formosa Plastics.[19] However, the results of the research plans commissioned by Formosa Plastics are mostly required to be kept confidential. The outside world would never know their specific research processes, methods, and data; let alone implement any scientific verifications or dialogues. On the other hand, Prof. C.C. Chan (who has been questioned by the parents) believed that science has already proven the high level of health risks brought by the SNCC to its neighboring communities. He stressed that he accepted the commissions from the NHRI and the Changhua County Government, and "all of the plans had undergone an average of 90-day strict review by the ethics committee of the NTU Hospital. The research funding sources as well as the research methods and processes were absolutely transparent and accountable."[20] C.C. Chan believed that it is unfair and unscientific to compare his published and publicly available research reports with those that did not undergo any peer review and were not published.

We have no way of knowing whether prior commissions by the industry would affect the scientists' test results. However, it is apparent that the inconsistent administrative decisions, which do not treat the knowledge production processes and inference methods rigorously but seek to rely on scientific data to provide simple solutions, would only generate more political as well as scientific disputes. When a "health exam" without strict research design, IRB review, or only limited to the SBCES schoolchildren is compared to a study with the scope of samples that covered the entire nation or region, was published on a regular basis, and had passed the IRB or peer reviews; how to interpret both findings and elaborate their different degree of reliance is rather political. The case showcases the government's passive action to reluctantly work on a more comprehensive research design for fact findings. As a result, political mobilization driven by economic reality took over the rational politics originally and was supposedly back up by rigorous scientific evidences.

11.4.3   Mud wrestling of science in SNCC's domination of local political economy

As mentioned, incomplete scientific investigations and inconsistent policymakings have intensified rather than resolved the school relocation controversies. A lack of delicate political decision-making processes and risk communications among different stakeholders further escalated the conflicts. For a community that heavily depends on the petrochemical giant for economic survival, even the best-intended policy can become an unbearable weight that further degrades the local resident's trust in "science" for its role of being solving problems.

Lin (2019) tried to explain why local people were not convinced by data and evidences that Prof. C.C. Chan's team has systematically gathered and scientifically presented against SNCC. In his essay "How Is the Local Made Not to Work? The Scientific War for and against the No.6 Naphtha Cracker Complex and the Geography of Trust of Mailiao Residents," he argued that the residents' challenges to the neutrality and validity of the anti-SNCC research are not simply ignorant or lacking understanding of science. He found that the way SNCC launched the "scientific war" against anti-SNCC research through claiming no regulation violation of monitoring data, or the way SNCC popularized and disseminated propaganda that attributed increasing cancer rates to individual habits rather than industrial pollutions, has been successful in re-interpreting bio-politics at local. Moreover, due to the governance failure at local that the Mailiao township development has heavily relied on SNCC's resources, the residents have lost trust in the local government and considered those who have been involved in the anti-SNCC campaign care nothing about residents' welfare but think only of their own self-interest.

Jobin (2020) further demonstrates why the fenceline community has intended to accept the chronic pollution with rare protests. He noticed that the rich company could pay generous grants for their "good neighbor's program": its hospital system offers free health care at Mailiao, the public elementary school students receive free meals, and the elderly benefit from home service care sponsored by the company. Besides, households receive monthly allocations with amounts that depend on the township of residency (p. 9). According to an investigative report, local township and village leaders in Mailiao have been involved in corruption scandals more frequently than elsewhere in Taiwan (Fang et al., 2019: 109–129). Tu et al. (2014) examined the local governance dilemma in dealing with SNCC's pollution and industrial safety accidents. They found that SNCC has been able to challenge the local regulators and manipulate the local politics through various tactics that include a strategic lawsuit against public participation (SLAPP), as well as setting feedback funds to pacify and buyout local politicians or dominate local political agenda in investigating SNCC (p. 85). As Jobin (2020) stated, SNCC has laid out a system of regular allocations and welfare patronage which is likely to prevent a large mobilization against itself. Money in this case is an effective tool for buying local support and diverting public attention away from its responsibility for the effects of chronic industrial pollutions.

In the SBCES relocation disputes, we do see from both the epidemiological study and the health check that the TdGA levels in the urines of SBCES schoolchildren were much higher. But on the environmental monitoring side, the EPA did not detect VCM or it was below the detection standard. Although the undetected VCM may have been subjected to the EPA's detection methods, timing of sampling, instrument's capacity, or factory's halting operation, the inconclusive scientific results have stirred a wave of local mobilization in favor of returning SBCES. As the disputes weakened the legitimacy of science, SNCC's domination of the local political economy, by shifting the health concerns to moral economic issues in which accusing anti-SNCC campaign as greedy profit-seekers, was even more prominent.

11.5 Conclusion

The SBCES relocation disputes demonstrate that science is not always able to clarify the facts or to verify a single causal relationship in an open environment.

Sometimes, it may instead intensify the conflicts or even mire the entire policymaking process if we ignore the political nature of science in the risk controversies. The disorganized interweaving of politics and science may lead to more disputes and significantly weaken the role of science in the policymaking process.

This case study indicates that the government has been passive to do a more comprehensive study for fact findings of the toxic exposure of local residents. I do not have a clear answer why the government has been reluctant to do so in order to protect the residents' health. However, it is apparent that SNCC has great political and economic influences at Mailiao through its resource domination and welfare patronage. Taiwan has long been characterized as a developmental state, as the central government played a dominant role in resource allocation and promoting industrial development. Although the anti-petrochemical movement has kept growing and successfully halted some giant petrochemical development plans in the twenty-first century, the government's regulatory management on the already developed petrochemical complex has approved little effective. Chou (2015) has criticized that the failure resulted from a regime of expert politics with hidden and delayed risk agenda. What is more demonstrated in this case study is that the governance failure generated social mistrust toward the government and created an opportunity for welfare patronage between the local community and the dominant industrial business. The state no longer plays a dominant role in leading industrial development. The increasing domination of SNCC on both local political influences and science claims may somehow silence the government as it creates tremendous political risks against the industrial interest.

Although the original school relocation policy appeared to be based on scientific evidence, the credibility and legitimacy of science-based decisions were deteriorated when the government agency commissioned other researchers to conduct a cursory investigation under the pressure of enterprise objections and local citizen protests. As the scope, the research subjects, the study methods, and even the IRB review requirements involved for this investigation were all different and incomparable with the original study, the new investigation results appeared to have no referential value and the science has been relegated to no more than a political tool. By casting the doubt on science in the policymaking process, the company has been able to advance its agenda to create political rumors that portray the anti-SNCC campaign as profit seekers.

The cases about how petrochemical companies influence production and interpretation of scientific data as well as risk perceptions of fence line communities are not limited to Taiwan. In the case of Louisiana State of the USA, the petrochemical fence line community members in Norco and New Sarpy were divided in response to companies' good neighbor initiatives (Ottinger, 2013). As Norco is often portrayed as part of Cancer Alley, some residents feel offended and are willing to accept industry scientists' or engineers' claims that Shell's emissions are not harmful to defend their town. In New Sarpy, residents felt struggling whether to secure investment and put forth images as nice places to live by accepting Orion's promise to maintain clean air, or to challenge the industry scientists' data in a way to push the company to improve its environmental performance. Similar to the school relocation disputes discussed in this chapter, company influences are powerful but subtle. As Ottinger (2013: 96) noticed, the petrochemical companies "rest their authority not only on the strength of their scientific knowledge and technical competence but on the ways in which their technical practices help 'build up' fence line communities and their images in the eyes of others."

In this chapter, I further argue that the policymakers had a lack of understanding of the dynamics of knowledge creation and application, which might be able to facilitate the policy goals if used well or deteriorate policy legitimacy if misused. For example, under the policy objectives of protecting children, the scientific investigation should not be confined to the health impacts of particular pollutants. Relocation is also not the only solution. To establish the credibility of the policymaking knowledge, the better approach would be to provide public reviews and debates of the knowledge production processes and inference methods; such as whether there were systematic deviations in the samples or the sampling processes, whether there were analysis instrument limits, or whether there was any information about the researcher's conflict of interest that must be disclosed. We may further ask what knowledge is required to achieve the policy goal and what kind of rigorous method is required to support the production of such knowledge in order to complete the science for solid policymaking.

The will to conduct more in-depth and rigorous scientific investigations involve the government's willingness to spend the resources and exercise its administrative powers in order for science to serve its purpose during the political policymaking process. The SBCES disputes reminded us that the policymakers should be sensitive about how to interweave science and politics appropriately; including the careful examination of the issue framing, research methods, and the conflicts of interests in the knowledge production process. Otherwise, the domination of national and local politics driven by the polluter's interests may marginalize the rational politics that highlight public interests. Perhaps the only way to avoid the misuse of science in the policymaking process is to develop the public capacity in reflecting and contextualizing the science in the local political economy.

Notes

1 Sarah Boseley (8 December 2006). Renowned cancer scientist was paid by chemical firm for 20 years. *The Guardian*. www.theguardian.com/science/2006/dec/08/smoking.frontpagenews

2 NTU RSPRC (2 September 2015). *Newsletter: Statement for the school return decision* (in Chinese). https://rsprc.ntu.edu.tw/zh-tw/m01-3/air-pollution/278-yunlin-children.html

3 MHW (22 August 2016). *Newsletter: MHW recommend preventive relocation to protect the health of the schoolchildren* (in Chinese). www.hpa.gov.tw/Pages/Detail.aspx?nodeid=1136&pid=3149

4 Zhigang Yeh (25 August 2016) The YCG Magistrate Lee will convene a contingency meeting (in Chinese). *Central News Agency.* www.taiwannews.com.tw/ch/news/2970961

5 Parents refused to relocation (in Chinese). (1 September 2016). *Appledaily.* https://tw.appledaily.com/life/20160901/3F5XTB5J6LAFY2B6J2WSJLY4SU/

6 Mengli Yang, et al. (7 September 2016). The environmental groups argued that the professor Lee should recuse himself due to conflict of interest (in Chinese). *The Chinatimes.* www.chinatimes.com/newspapers/20160907000426-260114?chdtv

7 HPA (7 February 2017) *Newsletter: The HPA holds a meeting of environmental monitoring and urine testing experts* (in Chinese). www.hpa.gov.tw/Pages/Detail.aspx?nodeid=1137&pid=7025

8 Liren Liu, et al. (8 June 2009). Local cancer rate has a "significant relationship" with Formosa Plastics operations (in Chinese). *The Liberty Times.* https://news.ltn.com.tw/news/focus/paper/309542

9 Shu Wei (22 August 2016). Formosa Plastics raises 4 major questions against the relocation decision. *Central News Agency.* www.taiwannews.com.tw/ch/news/2969564

10 Tianru Huang (10 September 2016). Public health issues become "scientific mud war" (in Chinese). *The Storm.* www.storm.mg/article/164393

11 Wenxin Zhang (14 August 2014). No hard evidence indicate the TdGA found in the school children may cause cancer in the future (in Chinese). *The Storm.* www.storm.mg/article/34675

12 HPA decides that urine test will be performed by NCKU (in Chinese). (3 February 2017). *Central News Agency.* www.chinatimes.com/realtimenews/20170203003922-260405?chdtv

13 Mianjie Yang (7 February 2017). The environmental groups questioned Professor Li 's test results (in Chinese). *The Liberty Times.* https://news.ltn.com.tw/news/life/breakingnews/1967126

14 Tianru Huang (10 February 2017). Less eating eggs, more health? (in Chinese). *The Storm.* www.storm.mg/article/221727

15 Formosa Plastics claims to be unable to prove that the pollution comes from VCM (in Chinese) (23 August 2016). *Appledaily.* https://tw.appledaily.com/property/20160823/ZYRX4JFPMKAOPVKNBD3QBT74GA/

16 The parents oppose the relocation decision from government (in Chinese) (25 August 2016). *United Daily News.* https://theme.udn.com/theme/story/6773/1918023

17 Formosa Plastics claims to be unable to prove that the pollution comes from VCM (in Chinese) (23 August 2016). *Appledaily.* https://tw.appledaily.com/property/20160823/ZYRX4JFPMKAOPVKNBD3QBT74GA/

18 HPA (6 September 2016) Newsletter: Urine tests for school children will be performed simultaneously by two research units (in Chinese). www.hpa.gov.tw/Pages/Detail.aspx?nodeid=1136&pid=3154

19 HPA decides that urine test will be performed by NCKU (in Chinese). (3 February 2017). *Central News Agency.* www.chinatimes.com/realtimenews/20170203003922-260405?chdtv

20 Tianru Huang (10 February 2016). Prof. Chan claim: My research can be reviewed by the public (in Chinese). *The Storm.* www.storm.mg/article/164394

References

Chan, Chang-Chuan, Yungling Leo Lee, and Shou-Hung Hung (2011). *Air pollution and health among residents near a petrochemical complex in Yunlin County:*

A cohort study. Environmental Protection Bureau, Yunlin County, Taiwan, 2009/ 07/20–2014/12/31. (Chinese).

Chou, Kuei-Tien (2015). From anti-pollution to climate change risk movement: Reshaping civic epistemology. *Sustainability*, 7(11), 14574–14596.

Davis, Devra (2002). *When smoke ran like water: Tales of environmental deception and the battle against pollution*. New York, NY: Basic Books.

Douglas, Heather (2005). Inserting the public into science. In Sabine Maasen and Peter Weingart (eds.), Democratization of expertise? Exploring novel forms of scientific advice in political decision-making. *Sociology of the Sciences*, 24, 153–169.

Fang, Hui-Cheng, Jong-Hsin Ho, Yu-You Lin, and Yi-Ting Chiang (2019). *A smoking island: Petrochemical Industry, our dangerous companion for fifty years*. Taipei: SpringHill (in Chinese).

Fisher, Elizabeth (2007). *Risk regulation and administrative constitution democratization of expertisenalism*. Oxford: Hart Pub.

Frickel, Scott, Sahra Gibbon, Jeff Howard, Joanna Kempner, Gwen Ottinger, and David Hess (2010). Undone science charting social movement and civil society challenges to research agenda setting. *Science technology Human Values*, 35(4), 444–473.

Hess, David J. (2007). *Alternative pathways in science and industry activism, innovation, and the environment in an era of globalization*. Cambridge, MA: The MIT Press.

Hinchliffe, Steve (2001). Indeterminacy in-decisions: Science, policy and politics in the BSE (Bovine Spongiform Encephalopathy) crisis. *Transactions of the Institute of British Geographers*, 26(2), 182–204.

Huang, Po-Chin, Li-Hsuan Liu, Ruei-Hao Shie, Chih-Hsin Tsai, Wei-Yen Liang, Chih-Wen Wang, Cheng-Hsien Tsai, Hung-Che Chiang, and Chang-Chuan Chan (2016, October) Assessment of urinary thiodiglycolic acid exposure in school-aged children in the vicinity of a petrochemical complex in central Taiwan. *Environmental Research*, 150, 566–572.

Jasanoff, Sheila (1990). *The fifth branch: Science advisers as policymakers*. Cambridge, MA: Harvard University Press.

Jasanoff, Sheila (2005). *Designs on nature: Science and democracy in Europe and the United States*. Princeton, NJ: Princeton University Press.

Jobin, Paul (2020). Our 'good neighbor' Formosa Plastics: petrochemical damage(s) and the meanings of money, *Environmental Sociology*, 1–14. https://doi.org/10.1080/ 23251042.2020.1803541

Lin, Hung-Yang (2019). How is the local made not to work? The scientific war for and against the No. 6 Naphtha Cracker complex and the geography of trust of Mailiao Residents. *The Journal of Geographical Science*, 93, 35–80 (in Chinese).

Michaels, David (2008). *Doubt is their product: How industry's assault on science threatens your health*. New York, NY: Oxford University Press.

Michaels, David, and Celeste Monforton (2005). Manufacturing uncertainty: Contested science and the protection of the public's health and environment. *American Journal of Public Health*, 95(S1), S39–S48.

Oreskes, Naomi, and Erik M. Conway (2010). *Merchants of doubt: How a handful of scientists obscured the truth on issues from tobacco smoke to global warming*. New York, NY: Bloomsbury Press.

Ottinger, Gwen (2010). Buckets of resistance: Standards and the effectiveness of citizen science. *Science, Technology, & Human Values*, 32(2), 244–270.

Ottinger, Gwen (2013). *Refining expertise: How responsible engineers subvert environmental justice challenges.* New York, NY: NYU Press.

Sarewitz, Daniel (2000). Science and environmental policy: An excess of objectivity, chapter in R. Frodeman (ed.). *Earth matters: The earth sciences, philosophy, and the claims of community.* Upper Saddle River, NJ: Prentice Hall, 79–98.

Smith, Kevin B., and Christopher W. Larimer (2016). *The public policy theory primer* (3rd ed.). Boulder, CO: Westview Press.

Tu, Wen-Ling, and Chia-Liang Shih (2014). The political role of scientific knowledge in the environmental impact assessment: Examining the health risk assessment disputes of the 6th Naphtha. *Taiwan Foundation for Democracy*, 11(2), 89–136 (in Chinese).

Tu, Wen-Ling, Chia-Liang Shih, and Wan-Ju Tsai (2014). The petrochemical lesson for the agricultural county: Reviewing Yunlin County's environmental supervision on the sixth naphtha cracking plant. *Journal of Taiwan Land Research*, 17(1), 59–90 (in Chinese).

12 Rethinking the sources of air pollution and urban policies in Hong Kong

Paulina PY Wong

12.1 Introduction and overview

To effectively manage air pollution through regulations and to improve public health advice, it is crucial to rethink and identify the sources of air pollution in Hong Kong. According to an analysis of the Environmental Protection Department of the Hong Kong Special Administrative Region (HKEPD) (EPD, 2017), over 80% of the high pollution days of Hong Kong in 2016 were caused by transboundary air pollution emitted from manufacturing activities in the Pearl River Delta (PRD) Region. However, a recently released report by the United Nations Environment Programme titled "A Review of 20 Years' Air Pollution Control in Beijing" covering the period from 1998 to the end of 2017, indicates a substantial improvement in air quality in Chinese cities, particularly Beijing, under its Clean Air Action Plan from 2013 to 2017. In addition to general public concern regarding the transboundary flow of pollutants from the PRD to Hong Kong, there is also concern about pollution from local sources. Do other fixed (such as public electricity generation and coal-fired power plants at Castle Peak and Lamma) and mobile local sources (such as road vehicles and marine vessels) play a role in the overall deterioration of air quality in Hong Kong? Although the air quality has undergone major improvements in recent years, roadside air pollution remains 70% higher than the target set by the World Health Organization (WHO) (Kao, 2018).

Air pollution has been widely recognized as a major global public health risk factor. It is the second leading cause of non-communicable diseases. Nitrogen dioxide (NO_2), carbon monoxide (CO) fine particulate matter ($PM_{2.5}$), respirable suspended particulates (PM_{10}), sulfur dioxide (SO_2), and ozone (O_3) are common urban air pollutants, primarily resulting from the emissions of power plants, motor vehicles, and other industrial activities. Both short-term and long-term exposure to these toxic air pollutants have been associated with serious health issues. Many studies have revealed the close relationship between mortality and air pollution, especially respiratory mortality in temperate countries. Notably, urban air pollutants, such as O_3 and SO_2, were found to be significantly related to all types of respiratory mortalities (Wong et al., 2002). In recent years, numerous epidemiologic studies have shown that long-term exposure to air

DOI: 10.4324/9781003211747-16

pollution will link to cardiovascular morbidity and mortality. Exposure to $PM_{2.5}$ was also found to be associated with type 2 diabetes and mortality, adverse reproductive outcomes, and neurologic effects. The close relationship between concentrations of PM_{10} and $PM_{2.5}$ and mortality and morbidity has been assured by the World Health Organization (2018) and were considered as Group 1 carcinogen: an agent proven to cause cancer in humans.

In recent years, more attention has been given to the impact of air pollutants on mental health illness and cognitive impairment (Helbich, 2018; Tzivian et al., 2015). These strong associations are particularly evident in vulnerable groups and populations with the greatest risk occurring in children, women, and the elderly (Wong et al., 2016; Yang et al., 2018) and mental illness (Braithwaite et al., 2019). Among all age groups, children are most vulnerable to the negative effects of air pollution because their bodies are still developing (Sunyer et al., 2017, Khreis et al., 2019).

Hence, to protect public health and improve public health advice, it is crucial to identify the sources of contributions and the causal factors of air pollution that lead to adverse physical and mental health effects. This chapter aims to present a grounded and neutral overview of the double exposure of transboundary air pollution and local emissions in Hong Kong, but the underlying argument involves the increase of local emissions, relaxed regulations, and standards. These major issues and challenges in the future demand a rethink. Additionally, the assessment methods and adaptive solutions, such as the emerging urban sensing technologies and applications that are leading to new possibilities and foster public participation will also be recommended and discussed in this chapter.

12.2 Background of regional development

The PRD economic zone (EZ) is one of China's leading economic regions and a major manufacturing conglomeration. The EZ embraces nine cities, namely Guangzhou (the provincial capital), Shenzhen, Foshan, Zhuhai, Jiangmen, Zhongshan, Dongguan, Huizhou, and Zhaoqing. In recent years, the inclusion of Hong Kong and Macao (Special Administrative Regions—SAR) has given rise to the Greater Bay Area (GBA). The GBA, an emerging megacity located at the southern coast of China and connecting to Hong Kong, is undergoing rapid economic transformation, particularly in digital and high-tech innovation. The new cross-border transport links (i.e. Zhuhai-Macau Bridge, high-speed rail network) have greatly enhanced cross-border travels, where most of the PRD cities are reachable within three hours from Hong Kong SAR. Geographic proximity has no doubt, contributed to strengthening not only the socioeconomic relationship but also mutual cooperation in environmental sustainability. As the world's fourth-largest financial center, Hong Kong's active involvement has greatly accelerated business opportunities in the GBA region where a mix of high-technology production, manufacturing capacity, and financial services offers a uniquely diversified economy worth US$1.5 trillion in GDP. The combined GDP of the GBA region is expected to triple by 2030 to US$4.6 trillion, surpassing the economic size of the New York Metropolitan Area ($2.2 trillion) (Jansen N., 2018).

The emergence of a broad range of industries, for example, garments, footwear, plastic products, electrical goods, transportation, logistics, financial services, etc., features the economic development of the GBA. The industrial cluster has become an integral part of the regional economy (Constitutional and Mainland Affairs Bureau, 2017). The GBA economy has benefited from the steady growth of the new industries. However, the rapid pace of industrial development has placed great strains on the infrastructure, environment, energy, and other public services in the GBA. The problem of environmental degradation, particularly cross-boundary air pollution, has become an emerging issue and major concern for both Hong Kong and PRD. Both regions were aware of the issues and determined to tackle them through a concerted effort under the "one country, two systems" principle.

Figure 12.1 A map of China's Pearl River Delta (PRD) and Greater Bay Area (GBA) and the corresponding socioeconomic development. Trade Development Council (2020)

12.3   Sources of air pollution and regulations

12.3.1   *Greater China and PRD pollution*

The PRD was one of the first regions to develop in China with its reforms and opening-up policies in the early 1980s. The PRD, a major manufacturing and industrial center in China, is known to have caused severe air pollution problems. The combustion processes of power plants, vessels, manufacturing, and industrial activities have produced a high concentration of SO_2, NO_2, CO, $PM_{2.5}$, PM_{10}, and indirectly O_3. The dramatic urban expansion, land use change, increasing trends of GDP, vehicle numbers, and electricity consumption were known to be the major key factors of degrading the air quality and affecting the climate of the region (Wang et al., 2007, Lin et al., 2011). In particular, electricity generation from coal-powered industries and households, which was mainly generated by fossil fuels and coal burning, was known to be one of the main sources of China's air pollution, reaching over 76% of energy consumption in China since 1990 (China Power Team, 2016). In the early years (the 1970s and 1980s), substantial emissions of SO_2 and NO_x into the atmosphere were then oxidized and transformed to sulfuric (H_2SO_4) and nitric acids (HNO_3) and transported across distances, causing acid rain problems on a regional scale (Larssen et al., 2006). As estimated by Lu et al. (2020), the influence of acid rain affected more than 30% of the total territory, particularly observed in the southern and southwestern regions in China. Zhang and McSaveney (2018) observed the average annual pH of Chongqing's rainfall varied between 4.3 and 5 (normal rain should measure around 5.6) during 1986 and 2014 at Chongqing, China. Acid rain became China's first serious cross-regional air pollution problem. It was similar to the "black triangle" in central Europe in the early 1980s (Larssen et al., 2006). Acid rain posed a severe threat to the environment, agricultural production, and to humans (Yang et al., 2002; Zhang and McSaveney, 2018).

The societal cost of acid rain in China was estimated to be 32 billion USD (Larssen et al., 2006). Reducing SO_2 emissions from coal combustion sectors has become more regulatory since the 1990s and the Two Control Zones (TCZ) policy was imposed in 1995 (Hao et al., 2000) to strictly limit the production and use of high-sulfur coal, and adoption of desulfurization units for newly built and renovated coal-fired power plants within the TCZ region. However, the policy was not well enforced, and worsened acid rain pollution and continued increase of SO_2 emissions were observed nationwide (China Meteorological Administration, 2014). Improvement was only observed after 2006 when the 11th Five Year Plan (FYP) enforced the control target for all regional leaders to adhere to (Jin et al., 2016). As described by Deng et al. (2008), many cities in China, especially the metropolitan Guangzhou, inevitably suffered from poor air and degraded visibility throughout the late twentieth century. Various literature and research have emerged since 2000 on air quality, visibility degradation, and especially the frequent occurrence of haze pollution (Wu et al., 2010). Citizens, tourists, and business travelers were all concerned about the notorious smog and worried about how it would affect them and their children? The haze pollution or smog was primarily caused by the integrated results of vehicle emissions under unfavorable weather conditions. Other factors, such as coal burning in neighboring regions, dust storms from the north and local construction dust, contributed as well.

To address the unprecedented issues and enable harder target control to be enforced, improved data monitoring and assessments to better support scientific study and evidence were crucial. Hence, the PRD Regional Air Quality Monitoring Network (RAQMN) was jointly established in 2003 by the Guangdong Provincial Environmental Monitoring Centre (GDEMC) and HKEPD and became operational in 2005. The RAQMN network was expanded to include the Macau region in 2014 and was renamed the "Quality Management Committee of Guangdong-Hong Kong-Macao Pearl River Delta Regional Air Quality Monitoring Network." The network currently consists of 23 automatic air quality monitoring stations (18 in Guangdong, 4 in Hong Kong, and 1 in Macau) across the PRD region to undertake quality management and dissemination of information for the Network (Guangdong-Hong Kong-Macao Pearl River Delta Regional Air Quality Monitoring Network, 2019).

All air quality monitoring stations are installed with instruments to measure the ambient concentrations of six major air pollutants: PM_{10}, $PM_{2.5}$, SO_2, NO_2, O_3, and CO. The U.S. National Ambient Air Quality Standards (NAAQS) are applied to the ambient air throughout the PRD region. The NAAQS enable clear-cut goals for air planners and regulators to achieve a "good" level of air quality, which is comparable to other measurements used throughout the world (Zhong et al., 2013). In 2018, a report of the monitoring results of the Guangdong-Hong Kong-Macao Pearl River Delta Regional Air Quality Monitoring Network was released. The results from 2006 to 2018 indicated the annual average of SO_2 had decreased nearly fourfold, from 47 µg/m³ to 9 µg/m³ and was in compliance with the concentration limits of SO_2 by NAAQS (60 µg/m³). The concentration levels of PM_{10} and NO_2 also showed an improvement. The annual average of PM_{10} met the national concentration limits of 70 µg/m³ since 2009, and NO_2 met the national concentration limits of 40 µg/m3 since 2011. Beginning in 2014, two parameters, $PM_{2.5}$ and CO, were added to the monitoring network and the concentration levels of $PM_{2.5}$ largely met the national annual average concentration limits of 35 µg/m³. CO also met the 24-hr average concentration limits of 4 µg/m³.

For the past few decades, there have been criticisms and doubts about the reliability of the air quality monitoring data released by China, due to the manipulation of data by local environmental protection. In April 2008, the US Embassy in Beijing started releasing hourly $PM_{2.5}$ readings via Twitter, from their embassy's roof-top air quality monitor located in Chaoyang District. Soon the consulates in Guangzhou, Shanghai, Chengdu, and Shenyang followed in November and December 2011, June 2012, and April 2013, respectively (StateAir, n.d.). Arguable standards and scrutiny of the integrity of the local air pollution data have left the authorities with serious credibility problems. The general public and scientists were skeptical about the local data. Finally, China's Ministry of Environmental Protection (MEP) began to report hourly PM_{10} and $PM_{2.5}$ data in 74 cities (which included the five cities with the US diplomatic posts) from January 2013, which was extended to 338 cities in January 2015 (Liang et al., 2016). Liang et al. (2016) also evaluated the data reliability of $PM_{2.5}$ in the five cities with US diplomatic posts and their nearby MEP sites between 2013 and 2015. The comparison outcome indicated that the air quality data from the U.S. posts and MEP stations were highly consistent in the five cities. Surprisingly, a declining trend was observed in 2015, possibly due to a decrease in energy consumption. Stoerk (2016) also suggested and indicated that the "misreporting of air quality data for Beijing has likely ended in 2012" upon the statistical validation with Benford's Law (Brown, 2005) and comparison with the US Embassy data. Though the concentration levels were still above the WHO recommendations, it was still encouraging to the community, with a "breakthrough" to open data initiative and public engagement.

Moreover, traditional studies and characterizations usually relied on sparse and fixed-site measurements, leading to incomplete spatial and temporal coverage. The data quality and integrity of some stations may lead to the underrepresentation of broader air quality in China. The advanced technological development in satellite-based remote sensing provides a useful alternative for estimating $PM_{2.5}$ concentration on a large spatial scale and enables better air pollution assessment on a city scale of the PRD region in the recent decade. Several studies have emerged in this discipline, providing more solid evidence to characterize the emissions and information to better combat the chronic air pollution problems in China (Li et al., 2015, Wong et al., 2017, Lin et al., 2018).

12.3.2   Local pollution in Hong Kong

According to the 2017 emission inventory report (EPD, 2019b) in Figure 12.2, one of the major sources of carbon emissions in Hong Kong came from public electricity generation, which contributes to serious SO_2 (43.3%), NOx (26.6%), PM_{10} (15.7%), and $PM_{2.5}$ (10.3%) problems. Hong Kong's buildings account for about 90% of the city's electricity usage. Over 60% of the carbon emissions are attributable to electricity generated for buildings. Other than electricity generation, road transportation is another major source of carbon emissions in Hong Kong. Road transport contributed to serious NOx, PM_{10}, $PM_{2.5}$, and CO problems but not SO_2. Since July 2010, the EPD has tightened the statutory motor vehicle diesel and unleaded petrol specifications to the Euro V level, which further tightens the cap on sulfur content from 0.005% to 0.001%. Road transport, therefore, is anticipated to contribute only a very small amount of SO_2 emissions. With the adoption of low-sulfur and ultra-low-sulfur fuel under the existing government policy, SO_2 is not a critical air pollutant of concern in Hong Kong.

However, the concentration levels of key air pollutants (i.e., NO_2, PM_{10}, $PM_{2.5}$, and CO) measured by the three roadside monitoring stations (managed by HKEPD—Figure 12.3) have consistently exceeded the local air quality objectives (AQOs) adopted by the HKEPD (EPD, 2005) and constantly breached the WHO's annual safety limit. Though the government had plans to phase out old diesel vehicles, minimal impact was achieved in lowering roadside pollution to safer levels in Hong Kong. This was due to fewer targeted policies and regulations on handling traffic congestion, a growing number of registered vehicles, and the slow pace in electrifying public transport. Many studies have also reported the severe traffic-related air pollution and health exposure impact in Hong Kong arising within the complex building morphologies and narrow urban street canyons resulting in a "street canyon" effect (Lee et al., 2017; Shi et al., 2018; Wong et al., 2019). Hence, better urban planning and design solutions combined with better transport management should be targeted for improved air quality outcomes as well as overall functionality and community well-being.

Figure 12.2 shows that marine navigation has contributed significantly to all pollution especially, SO_2 (51.7%), $PM_{2.5}$ (40.7%), NO_x (37.2%), and PM_{10} (34.1%). Hong Kong is one of the world's largest shipping hubs and one of the busiest and most efficient international container ports in the world. It handled 18.3 million TEUs (twenty-foot equivalent units) of containers in 2019. The port provided about 300 container liner services per week connecting to around 420 destinations worldwide. The Kwai Chung-Tsing Yi Container Terminal (Figure 12.3) is among the most important port/container terminals in the world and accounts for over 77% of port containers in Hong Kong. More than 10,000 ocean-going cargo vessels arrive at the terminals each year. The terminal serves mainly as a gateway port for South China cargo and as a trans-shipment hub (Transport and Housing Bureau, 2018). Hence, the active shipping business contributes to the serious SO_2 and NO_2 problems in Hong Kong.

In 2015, in the Kwai Chung area, the 24-h SO_2 concentration exceeded the WHO's limit (20 µg/m³) for 108 days (Environment Bureau, 2017a). The SO_2 emissions from ocean-going vessels (OGV) contributed 40% of the total SO_2 emissions in Hong Kong in 2014 (Environmental Protection Department, 2016). For the annual averages of NO_2 concentration from 2012 to 2014, the Kwai Chung Region continuously exceeded the annual limit of at least 20 µg/m³ (Environment Bureau, 2015). The University of Hong Kong's School of Public Health also estimated that pollution in the Kwai Chung area would cause respiratory problems to an additional 400 children aged 8–12, which is the most found in Hong Kong (Wong et al., 1998; Hora, 2010).

The OGVs and marine-related pollutants are a major source of air pollution in Hong Kong. Though 60% of the emissions come from container ships/cargo vessels, cruise liners that disembarked at the cruise terminals of Kai Tak and Ocean Terminal of Hong Kong, were other contributors to air pollution. As described by Martin Cresswell, technical director of the Hong Kong Shipowners Association, "*While idling, a cruise liner such as the World Dream emits as much SO_2 as would 25,000 local diesel buses*" (HSN, 2019). Cruise ships are required to continue running their engines and power everything up (i.e., entertainment, restaurants, rooms) while disembarking at the cruise terminal. This alone is estimated to burn up more than ten times the fuel of small container ships (HSN, 2019).

As illustrated in Figure 12.3, the locations of the major port facilities, shipping channels, and cruise terminals were all situated close to the urban population, in residential, business areas, and parks of Hong Kong. According to the 2016 Census, there are over 520,000 residents living in the Kwai Tsing District (Census, 2018) and about 3.8 million people in nearby districts (Lau et al., 2004). People living and working in these areas are likely to have constant exposure to much higher concentrations of air pollutants, leading to adverse health consequences. Green groups (i.e., Clean Air Network, Greenpeace), academic institutions, and local residents have been making great efforts for the last two decades, hoping to influence the government to better control marine emissions, reduce the public health risk, and tackle the city's growing pollution problem.

In 2010, the world's first industry-led voluntary, unsubsidized, at-berth fuel-switching initiative, known as the *Fair Winds Charter* was conducted in Hong Kong. It was formulated by the Civic Exchange, a Hong Kong-based public policy think tank, and the Hong Kong Line Shipping Association, and supported by the University of Science and Technology to monitor and assess the environmental benefits (Ng, 2018). A total of 17 companies signed the charter and agreed to voluntarily switch to burning cleaner fuels (0.5% sulfur content or lower) instead of high-sulfur fuels while their vessels were berthed in Hong Kong. The voluntary fuel switch led to a reduction of 4.9% SO_2 and 3.6% PM_{10} from the shipping industry (EPD, 2013). The Kwai Chung air-quality-monitoring station, in the area of the container terminals/ports, has also recorded approximately 60% reduction in average SO_2 concentrations for the following three years, compared to 2014 (HSN, 2019).

The charter not only demonstrated a great role model by engaging the government to influence policymaking and future emissions regulations, but also laid a good foundation of "science-to-policy" practice, showing how scientific research could successfully support policy change (Ng, 2018; HKUST, n.d.). The Air Pollution Control regulation for OGVs (fuel at Berth) eventually became effective in 2015. In early 2019, a regulation, "Fuel for Vessels," was also implemented in Hong Kong requiring all vessels in Hong Kong waters to use fuel with a sulfur content of no more than 0.5% (previously 3.5%). Though there is a consensus that the government's marine and road sulfur caps have significantly reduced SO_2 pollution, the regulations have had less effect on other pollutants. Scientists and green groups in Hong Kong are particularly concerned about risks from the NOx and PM that are emitted by ship exhaust. Moreover, transboundary flows of these pollutants from the PRD to Hong Kong were also a concern, since Shenzhen and Nansha (Guangzhou), located in the GBD, were two of the world's biggest ports, accounting for 39% of China's export cargo. The mutual understanding and combined effort from China and PRD to join the clean-up action in Hong Kong are crucial.

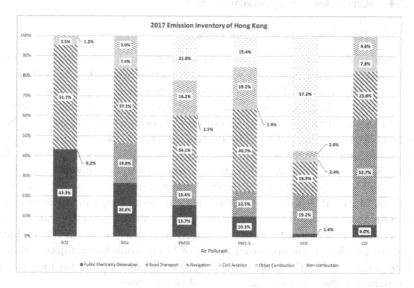

Figure 12.2 The emission inventory (percentage share) of Hong Kong during 2017
Source: (EPD, 2019b)

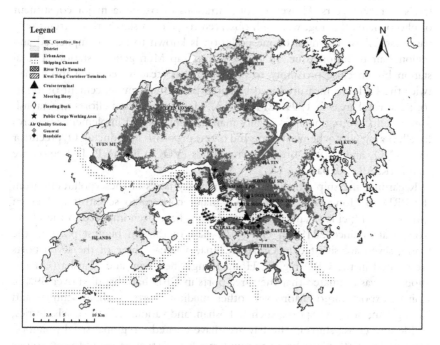

Figure 12.3 The distribution map of Hong Kong port facilities, major shipping channels, and cruise terminals which were all situated closely to the urbanized areas (shaded in dark grey). The location of three roadside and 16 general air quality stations managed by HKEPD were also included (EPD, 2019b).

12.4   Rethinking responsibilities

12.4.1   *Transboundary pollution impact*

Previous studies have reported that various air pollutants (i.e., $PM_{2.5}$ and O_3) could travel on intercontinental scales and toward downwind regions (Uno et al., 2017; Hou et al., 2019; Liu et al., 2020). Transboundary pollution impacts from PRD to Hong Kong were found to contribute the most in winter and fall due to prevailing northerly wind direction. According to the report published by EPD (2018), in the winter in Hong Kong, with prevailing north and northeast wind direction, as much as 77% of the dirty air/dust comes from China. In Hong Kong, districts in western areas were observed to be the worst, especially the Tuen Mun region, which was the most polluted district for the fifth consecutive year since 2014. Lin et al. (2018) also observed long-range transport of air pollutants (especially $PM_{2.5}$) from the center of the PRD region and noticed higher $PM_{2.5}$ concentrations in the northwest areas of Hong Kong. They added that this was due to the northwest region of Hong Kong being adjacent to the central area of the PRD region, making it the most susceptible to the long-range transport of pollutants from satellite-based observations.

With the joint efforts in cutting emissions in the PRD Region, the ambient levels of NO_2, SO_2, PM_{10}, and $PM_{2.5}$ have seen a reduction in concentration levels in recent years. However, concentrations of ozone, a major constituent of photochemical smog, were still observed to be on a slowly rising trend over past years. Photochemical ozone pollution is known to be a complicated and regional air pollution issue. In Hong Kong, Tap Mun general station, a rural station but in close proximity to Shenzhen, has steadily recorded more than twice the O_3 levels measured in urban areas since the commencement of monitoring at rural areas in 1999 (EPD, 2015). Hence, concerted efforts to establish a more effective regional air quality management plan to implement measures for alleviating photochemical smog and ozone problems by reducing O_3 precursor levels (NO_2 and volatile organic compounds (VOC)) in the PRD Region are vital (HKSAR, 2020a).

Regarding the ship emissions, transboundary flows of these pollutants from the PRD to Hong Kong were also a major concern to scientists and green groups. Even fuel laws and stricter regulations have been imposed in the Hong Kong waters since 2019 on OGVs. In view of the ten biggest ports in the world, seven are situated in Greater China, whereas, three of the seven ports are located in the GBD region: Hong Kong, Shenzhen, and Nansha (Guangzhou). It was estimated that the three ports in GBD have handled over 39% of China's export cargo. Moreover, other medium ports, such as Zhuhai and Zhongshan; busy river ports, such as Foshan; and various fishing vessels, ferries, and leisure cruise ships in the bay area have created a regional-sized pollution blanket. In 2010, Civic Exchange undertook an extensive study to investigate the health effects of marine emissions in both regions. The report indicated that the annual number of deaths attributable to SO_2 emissions in the PRD was

519, of which 385 (or 74%) took place in Hong Kong (HSN, 2019). The lead author of the report, Mr. Simon Ng Ka-wing, the director of policy and research at the Business Environmental Council added:

> We calculated deaths from SO_2 only, but the total premature deaths from all shipping-derived air pollution would be at least double that [519] figure. Most of the ships accessing the Greater Bay Area ports do so via Hong Kong [waters] because it's more convenient, but Hong Kong is suffering and not deriving any economic benefit as they pass through.

Hence, the problems are expected to worsen as the GBD area continues to grow (HSN, 2019). With the support of the US Consulate in Guangzhou, sharing and communication between Civic Exchange and Shenzhen Municipality was established in 2011 to promote the pioneer project in Hong Kong and the concept of Emission Control Area (ECA). Soon in 2014, Shenzhen proactively introduced similar voluntary incentive campaigns at the Shenzhen ports and laid a good foundation for other PRD ports to join such clean-up actions. In December 2015, China's Ministry of Transport began requiring OGVs to use low-sulfur fuel (0.5%) in 11 major ports. In January 2019, the restriction was extended to three new domestic emission control areas (DECAs), which covered the GBA. Since then, all OGVs entering the DECAs will be required to use low-sulfur fuel. The significant development in China was welcomed by the HKSAR government and nearby cities/countries, which succeeded in enormous emission reduction and health benefits to the society (Ng, 2018).

12.4.2 Local impact and arguable responsibility

The various marine and road sulfur regulations implemented in recent decades have proven to have significantly reduced the SO_2 pollution in Hong Kong. Downward trends in other pollutant concentration levels were also observed due to a series of emission control policies. However, with the continued increase of vehicular fleets, traffic congestion, and the growth of the shipping industry in Hong Kong, concerns have arisen regarding regulations and policies that have had less effect on other pollutants, such as NO_x, NO_2, $PM_{2.5}$. Green groups and scientists in Hong Kong are particularly concerned about the risks from the NO_x and PM that are emitted in the exhaust from ships, intensive energy consumption, and roadside traffic.

According to an air pollution research conducted by a local university for the Civic Exchange, in 2007, the primary influence of regional sources and local sources on Hong Kong's air pollution were respectively around 36% (132 days) and 53% (192 days) at the time (HKUST & Civic Exchange, 2007). The Clean Air Network also noted that the PRD region's contribution to the air quality in Hong Kong was relatively low (Clean Air Network, 2017a). Lu et al. (2016) showed that the average local contributions of NO_x went from 42.9% to 60.8%, while the regional contribution (PRD region) contributed from 35.5% to 41.3%

on average. This indicates that the local region has a greater role in Hong Kong's air pollution than the PRD region, in contrast to the typically held opinion.

As illustrated in Figure 12.2, attention should be given to public electricity generation, which is the city's second-largest contributor to local air pollution. The emissions were primarily attributed to coal-fired power plants and electricity generated for city buildings. Though a mandated emissions cap has been introduced, there were no substantial large-scale planning or investments in renewable energy generation in the short term. As mentioned in the Environmental Bureau (2017b) Report, "Hong Kong does not have favourable conditions for large-scale commercialised RE (renewable energy) generation." Thus far, only a few small pilot projects were under way such as the photovoltaic (PV) systems installed in small or village houses, and the launch of Feed-in Tariff Scheme and floating PV systems at the city's reservoirs. In the long term, with the growing population, the public electricity generation and building electricity sectors would become a burden to society.

Other than emissions from ships and electricity generation, the third focus of attention should go to the road transport sector. Hong Kong, as a densely packed city filled with high-rises and low ventilation in urban areas, has created a severe street canyon effect as demonstrated in the empirical case study (HKD3D). The overwhelming vehicles on the road, especially public transportation, consist of more than 95% that are still powered by gasoline and fossil fuels. Though vehicle emissions standards and incentives for the replacement of diesel vehicles were enforced, the roadside AQ monitoring stations in Hong Kong often register poor and unhealthy traffic-related air pollution levels. Despite this, the adoption rate of clean vehicles in the city was comparatively low, particularly in the public transport sector. So far, fewer than 200 vehicles (i.e., electric buses, light goods vehicles, light buses, taxis) were on trial, supported by the HKSAR 28th Batch of Pilot Green Transport Fund (HKSAR, 2020b). By comparison, in Shenzhen, the public transport switched to a 100% electric bus fleet by the end of 2018. There are now approximately 16,000 electric buses and soon all 22,000 taxis will also be electric. The Shenzhen Bus Group estimated that they will be able to conserve 160,000 tonnes of coal per year and reduce annual CO_2 emission by 440,000 tons with the halved fuel bill (Keegan, 2018). Though there might be limitations in battery power and economic costs, Hong Kong should, nevertheless, accelerate their drive toward electrified public transit and tackle the growing burden in the road transport sector. In addition, the impact on climate change and temperature rise would worsen the street canyon effect in urban areas of Hong Kong.

12.4.3 An empirical case study: HKD3D

This HKD3D (A Dynamic Three-Dimensional Exposure Model for Hong Kong) project led by King's College of London and supported by the University of Hong Kong and HKEPD (funded by Health Effects Institute) developed a dynamic three-dimensional model to estimate the exposure to traffic-related air pollution in Hong Kong (Barratt et al., 2018; Wong et al., 2019). The project has conducted vertical outdoor monitoring to measure $PM_{2.5}$, black carbon (BC), and other traffic-related air pollutants at six strategic locations for two weeks in summer and winter during 2014 and 2015. Continuous 24-hour measurements were carried out at four different heights of a residential building and on both sides of a street canyon. Subsequently, the $PM_{2.5}$ and BC data were used to develop three-dimensional vertical models for further health impact analysis. Figure 12.4 illustrates the seasonal monitoring outcome of $PM_{2.5}$ and BC at a "deep" urban street canyon situated at North Point in Hong Kong (aspect ratio = 3.6).

$PM_{2.5}$ was known to be a pollutant that can transport a long distance. BC is a primary component of local $PM_{2.5}$, usually influenced by local sources (Zhang et al., 2018; Wong et al., 2019). As illustrated in Figure 12.4a, the mean diurnal trend of BC in both seasons was quite similar. Specifically, the two peaks during two heavy traffic times (morning and evening rush hour) were more significant in winter due to the street canyon effect. The low ventilation due to prevailing northerly wind direction (street oriented in east-west direction) coupled with the closely built high-rises and heavy traffic have trapped the air pollutants and escalated the traffic-related air pollution within the deep street canyon. The street canyon effect is often cited as one of the factors in Hong Kong's worsening air pollution (Wong et al., 2019). In the summer period, the prevailing easterly wind direction better ventilated the BC within the street canyon. Similar outcomes were observed in the vertical profile with lower floors experiencing similar BC concentrations in both seasons, which denoted the BC mainly originated from local emissions with a lower regional contribution (Barratt et al., 2018).

The $PM_{2.5}$ outcome (Figure 12.4b) showed strong seasonal differences in mean concentrations throughout the day, denoting a strong regional contribution. The shift to northerly winds in winter leads to transboundary air pollution transport from PRD to Hong Kong. This produced an increment of between 20 and 35 $\mu g/m^3$ over summer concentrations. The $PM_{2.5}$ measurements were in compliance with the HKAQO limit (75 $\mu g/m^3$) but the winter measurements exceeded the WHO AOQ concentration limit (25 $\mu g/m^3$) throughout the day, denoting a health risk to the public. However, the contribution of BC to $PM_{2.5}$ in this street canyon reached approximately 5% and 9% in summer and winter respectively, which illustrated that BC plays an important role in local PM2.5. According to Janssen et al. (2012), severe BC pollution has also been reported with scientific concerns due to its negative effects on public health.

12.4.4 Arguable HKAQO limits and compliance with WHO guidelines

In Hong Kong, HKEPD relied on the AQOs to implement specific air quality control and green policies. The Air Pollution Control Ordinance (Cap. 311) establishes these objectives, which include the concentration limits for

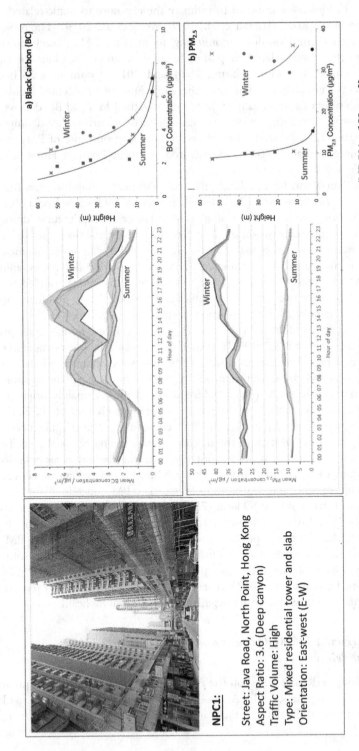

Figure 12.4 The mean diurnal and vertical variations of (a) BC and (b) $PM_{2.5}$ at an urban street canyon (NPC1) of Hong Kong

Source: (Modified from: Barratt et al., 2018)

Table 12.1 The AQO of Hong Kong, compared to WHO guidelines and the compliance status with HKAQO based on air pollution data in 2019

Air Pollutants	Averaging time	WHO Concentration limit ($\mu g/m^3$)	Current HK AQO		Compliance Status (2019)	
			Concentration limit ($\mu g/m^3$)	Number of exceedances allowed	General Stations	Roadside Stations
Sulphur Dioxide (SO$_2$)	10-minute	500	500	3	Yes	Yes
	24-hour	20	125 [a]	3	Yes	Yes
Respirable suspended particulates (PM$_{10}$)	24-hour	50	100	9	Yes	Yes
	Annual	20	50	—	Yes	Yes
Fine suspended particulates (PM$_{2.5}$)	24-hour	25	75 [b]	9 [b]	Yes	Yes
	Annual	10	35 [b]	—	Yes	Yes
Nitrogen dioxide (NO$_2$)	1-hour	200	200	18	Yes	No
	Annual	40	40	—	No	No
Ozone (O$_3$)	8-hour	100	160	9	No	Yes

Source: (Environmental Bureau, 2019 and Clean Air Network, 2018)

Note:

[a] *Proposed tightening of 24-hour average of SO$_2$ from 125 to 50 $\mu g/m^3$.*

[b] *Proposed tightening of annual average PM2.5 from 35 to 25 $\mu g/m^3$ and 24-hour average PM 2.5 from 75 to 50 $\mu g/m^3$, but with number of exceedances allowed increasing from 9 to 35 times.*

hazardous air pollutants. A review will be proposed at least once every five years to maintain an acceptable level of air quality to safeguard the health and well-being of the community. The AQOs are benchmarked against the targets of the WHO's air quality guidelines and are comparable to the air quality standards adopted by the European Union and the USA (Environment Protection Department, 2005). In other words, these AQOs will directly impact the public health of Hong Kong. Thach et al. (2018) proposed novel methods to construct a new air quality index in order to predict the health effects in Hong Kong. Li et al. (2018) proposed a modified analysis-forecast system for predicting the air quality index. This forecasting system consists of a complex data analyzing process and aims at forecasting the air quality index hourly.

However, this objective continually faces challenges by the green groups and scientists regarding its loose standards of air pollutant emissions (Creery, 2019) and non-compliance with the WHO guidelines (Clean Air Network, 2018). Table 12.1 illustrates the difference of AQO guidelines being adopted in Hong Kong for each pollutant compared with WHO guidelines since 2014. In July 2019, the HKEPD launched a three-month public consultation on the review of AQOs (Environmental Bureau, 2019). Though there was a proposed tightening of the objectives for SO_2 and $PM_{2.5}$ based on the WHO interim target, the number of exceedances allowed annually for $PM_{2.5}$ have disappointingly risen from the current 9 to 35 times the proposed number. Green groups and scientists expressed disappointment toward the new recommendations and argued that the proposed tightening of the annual $PM_{2.5}$ target made no improvements to the air quality in the near future, since Hong Kong on average was within the proposed annual target ($25\mu g/m^3$) by referring to the daily pollution data in 2018. Only one station (Tuen Mun) had slightly exceeded the proposed annual target (Tuen Mun's average was $26\mu g/m^3$); the average result across Hong Kong was $19.8\mu g/m^3$, which was well within the new annual target of $25\mu g/m^3$ as proposed by the new law. Hence, it is doubtful if the government will consider the concerns of public health a higher priority (Clean Air Network, 2018).

12.5 Recommendations

12.5.1 Strengthening cooperation at all levels

Considering that China has achieved significant improvement in recent years, is Hong Kong still primarily suffering from transboundary pollution originating in China? Overlooking the need to reduce local sources of air pollution might have exacerbated the problem. Hence, understanding the joint responsibilities and strengthening regional cooperation on tackling air pollution as a whole (not targeting individual pollutants) will help us gain greater awareness of the pollution hotspots and understand how they have impacted public health and contributed to episodes of regional haze.

As indicated by Zhong et al. (2013), no individual cities can successfully tackle air pollution and achieve better air quality on their own. A long history of regional cooperation between the PRD region and Hong Kong on various environmental issues has transpired since the early 2000s. A joint prevention and control system for air pollution on regulations, standards, and policies has been solidified (MEP, 2010). Improved air quality in 2010 as compared with 1997 has demonstrated the value of the joint effort, although greater effort would be required to significantly achieve clean air. Particularly, the ambient and roadside ozone (O_3) is still on a rising trend, it was forecast that by 2025, the concentration of O_3 in Hong Kong will continue to rise and exceed the relevant AQOs. To effectively tackle the local O_3 pollution, reducing the precursors of O_3 would be vital. However, regional cooperation on conducting more joint studies and stricter emission targets played a key role in tackling ambient O_3, as high regional background was recorded from the Tap Mun station (HKSAR, 2019a).

Hou et al. (2019) also revealed that a significant impact of transboundary air pollution and effectiveness of integrated measures between cities will be higher than an individual pollutant control strategy in improving the $PM_{2.5}$ levels. Hence, regional cooperation should continuously be promoted to address the problem of a lack of consensus on the issues among different countries (Kim, 2007). Regional cooperation under a consensus enables source countries with less capability to obtain assistance from others in order to take appropriate and effective actions.

12.5.2 Reliable assessment and representative exposure estimation for public health study

In most cities, air pollution and various pollutants are traditionally monitored routinely by fixed air quality monitoring stations located away from the roadside, and often the data do not sufficiently capture the small-scale spatial variability of air pollutants within urban environments and neighborhoods (Schneider et al., 2017). Recent studies have empirically confirmed that the concentrations of air pollutants could vary greatly within a small area or even just a few meters apart. Other factors, such as wind direction and wind speed, could make a notable difference as well. Reliability of health outcome estimations would, therefore, be reduced due to the low representativeness of data and biased responses. To effectively distinguish and analyze transboundary air pollution, appropriate methodologies, data quality assessment, and comparable air quality standards must play an important role. First, access to reliable air quality data is fundamental to the assessment. However, is the sparse network of air quality monitoring stations in Hong Kong sufficient to provide adequate data for public policy communication? Does the placement of these monitoring stations (13 general stations and 3 roadside stations) reflect the real pollution levels? (EPD, 2019) This monitoring network was certified by the Hong Kong Laboratory Accreditation Scheme. However, green groups and scientists (Clean Air Network,

244 *Paulina PY Wong*

2017b; Nip, 2017; Wong et al., 2019) argued that the current number of monitoring stations, particularly with only three roadside stations, was unable to collect detailed local data on air quality. Second, the monitoring stations were located in park-like settings, away from the urban area, and measurements collected from these stations could not reflect accurate pollution levels to which the residents are exposed. Hence, the air pollution level in Hong Kong might have been underestimated and its representation may be debated.

What are the quality measurement assessments that are needed to improve reliability? Satellite observations have been introduced as an alternative approach for estimating $PM_{2.5}$ concentration on a large spatial and temporal scale of a city. However, the measurement of other pollutants is limited and may not be ideal for capturing the small-scale spatial variability of air pollutants or a street-level study. Hence, should emerging urban sensing technologies and applications be used to enhance existing urban and health studies and to meet the challenges for sustainable urban development? In response to the need for up-to-date information on air quality to enhance public awareness, many researchers (Schneider et al., 2017; Hasenfratz et al., 2015; Moltchanov et al., 2015; Miskell et al., 2017) have promoted the use of low-cost microsensors and modeling methods to present real-time air quality on mobile apps and websites. Various novel approaches have also emerged for analyzing and presenting the spatial change of air quality and health-related risks in an urban setting.

Moreover, following the Triple Helix model (Etzkowitz & Leydesdorff, 1997) by nurturing collaboration about knowledge and technology transfer between universities and industrial sectors and government. Research and Development will enable technological advancement and fit-for-purpose microsensors, predictive models, and platforms to be manufactured and adopted by the government for large-scale/long-term deployment within highly concentrated urban environments. Chen et al. (2017) and Lee et al. (2019) demonstrated similar collaborations in Taiwan (i.e., The Edimax AirBox project). More than 2,500 sensors have been deployed across Taiwan for complementary data collection. The "real-time" air quality data were also opened and shared with the general public as raw data for further technological research and innovation. In recent years, Penta Helix Collaboration (extended from the Triple Helix model) was widely discussed to include stakeholders such as non-governmental institutions, civil society, and/or media to enhance innovation and smart city development (Sudiana, 2020).

12.5.3 *Greater data transparency and public participation*

The emerging urban sensing technologies, applications (i.e., personal air monitoring sensors) and "real-time" air quality data are leading to new possibilities in public health and community education. These platforms and data will not only be viable for producing reliable information about personal exposure to environmental stressors, but its "real-time" capabilities will also feature a greater data transparency, such as US embassy data from China.

In September 2018, the Hong Kong government had formulated a new policy on opening up government data. Annual open data plans have been published to facilitate the development of smart city and big data. Since late 2019, thousands of datasets have been gradually available to the public via the Public Sector Information Portal for free viewing and use (HKSAR, 2019b). The past 24-hour pollutant concentration of individual air quality monitoring stations is available every hour, but there is limited availability on past records and is only released as air pollution index instead of individual pollutant concentration.

Citizens that can access and capitalize on localized air quality and open data are believed to empower community education and citizen science research projects (Clements et al., 2017) and lay a strong foundation for civic engagement on big social issues and negative air quality health impact to the public. These kinds of local-specific measures will not only increase public awareness and increase social cohesion but will also improve the quality and quantity of the collected exposure data to better improve the air quality in both local and regional areas. Clean Air Network had been actively running Community Monitoring Projects with schools and communities around Hong Kong (Clean Air Network, 2017b). Ultimately, it will increase the effectiveness of air pollution controls and better advise the government to take appropriate preventative and mitigative actions to maximize the quality of life.

12.6 Concluding remarks

Air pollution has been widely recognized as a major global public health risk factor and the dangerous air pollutants were rated as a Group 1 Carcinogen by WHO. For the last few decades, the rapid development of China and PRD has resulted in severe air pollution. Air quality data and an enormous amount of studies have noted transboundary pollutants indeed coming from across the border and impacting Hong Kong, particularly during cool seasons with northerly wind direction. However, effective regional cooperation with strict regulations and emissions standards were enforced in recent decades, the air quality in both Hong Kong and PRD had undergone major improvements. Studies based on satellites and ground-based sensors have noted declines in key pollutants in recent years in many areas, notably the Beijing-Tianjin Metropolitan Area and the PRD (Voiland, 2018). Despite this, the levels of air pollution in China and Hong Kong still remain much higher than the AQO target set by the WHO, which required continued attention and regional cooperation.

The HKSAR government has also been criticized by the green groups and scientists for doing too little to clean up the air. "The government has been careless and intransigent on this issue," said Prof. Anthony Hedley, a professor in the School of Public Health of the University of Hong Kong, adding, "the government has de facto traded off child health and individual life expectancy against other vested interests" (Hora, 2010). Hopefully, the HKSAR government will be more responsive to local problems and apply stricter laws, policies, and regulations on air pollution control by implementing better air quality data assessment for evidence-based environmental policymaking (Storek, 2016), realizing the co-benefits of reducing air pollution for climate change mitigation, and, thus, provide better protection for the society and vulnerable populations.

Acknowledgment

The author would like to thank Prof. Joshua Mok for introducing her to Prof. Kuei-Tien Chou and the research team in Taiwan. The author is very grateful to Prof. PC Lai for her valuable comments and suggestions on the manuscript, and also Ms. Wenhui Cai for her assistance with manuscript preparation.

References

Barratt, B., Lee, M., Wong, P.Y.P., Tang, R., Tsui, T.H., Choi, C., Cheng, W., Yang, Y., Lai, P.C., Tian, L., Thach, T.Q., Allen, R., & Brauer, M. (2018) A dynamic three-dimensional exposure model for Hong Kong. *Research Report 194*. Boston, MA: Health Effects Institute.

Braithwaite, I., Zhang, S., Kirkbride, J.B., Osborn, D.P.J., Hayes, J.F. (2019) Air Pollution (Particulate Matter) Exposure and Associations with Depression, Anxiety, Bipolar, Psychosis and Suicide Risk: A Systematic Review and Meta-Analysis. *Environmental Health Perspectives*, 127(12), CID: 126002.

Brown, R.J.C. (2005) Benford's Law and the Screening of Analytical Data: The Case of Pollutant Concentrations in Ambient Air. *Analyst*, 130, 1280–1285.

Census (2018) *Population By-census 2016—By-census Results, Census and Statistic Department of HKSAR*. Retrieved from www.bycensus2016.gov.hk/en/

Chen, L.J., Ho, Y.H., Lee, H.C., Wu, H.C., Liu, M.H., Hsieh, H.H., Huang, Y.T., & Lung, S.C.C. (2017) An Open Framework for Participatory PM2.5 Monitoring in Smart Cities. *IEEE Access*, 5, 14441–14454.

China Meteorological Administration (2014) *Annual Report of Acid Rain Monitoring in China*. China Meteorological Administration (CMA), Beijing.

China Power Team (2016) *Is Air Quality in China a Social Problem? By China Power*. Retrieved from https://chinapower.csis.org/air-quality

Clean Air Network (2017a) *Where do Hong Kong's Air Pollutants Come From?* Retrieved from www.hongkongcan.org/hk/article/where-does-hong-kong-air-pollution-come-from/

Clean Air Network (2017b) *Community Monitoring Project*. Retrieved from www.hongkongcan.org/hk/article/community-monitoring-project/

Clean Air Network (2018) AQO Review Coalition Opposing EPD's Recommendations on the Air Quality Objectives Review. *Clean Air Network*. Retrieved from www.hongkongcan.org/hk/article/recommendations-second-aqo-released/

Clements, A.L., Griswold, W.G., Rs, A., Johnston, J.E., Herting, M.M., Thorson, J., & Hannigan, M. (2017) Low-Cost Air Quality Monitoring Tools: From Research to Practice (A Workshop Summary). *Sensors (Basel, Switzerland)*, 17(11), 2478.

Constitutional and Mainland Affairs Bureau (2017) *An Overview of the Greater Pearl River Delta*. Retrieved from www.cmab.gov.hk/tc/images/issues/ar0405/F_GPRD_Overview.pdf

Creery, J. (2019, September 8) Clean Air Network: Hong Kong Air Pollution NGO Urges Gov't to Prioritise Public Health. *Hong Kong Free Press*. Retrieved from www.hongkongfp.com/2019/09/08/clean-air-network-hong-kong-air-pollution-ngo-urges-govt-prioritise-public-health/

Deng, X., Tie, X., Zhou, X., Wu, D., Zhong, L., Tan, H., Li, F., Huang, X., Bi, X., & Deng, T. (2008) Effects of Southeast Asia Biomass Burning on Aerosols and Ozone Concentrations Over the Pearl River Delta (PRD) Region. *Atmospheric Environment*, 42(36), 8493–8501.

Environmental Protection Department (2013) *Hong Kong Air Pollutant Emission Inventory for 2011*. Retrieved from www.info.gov.hk/gia/general/201303/20/P201303200307.htm

Environment Bureau (2015) *Annual Averages of Air Pollutants (in µg/m3) Recorded by General Air Quality Monitoring Stations on Both Sides of the Victoria Harbour between 2012 and 2014*. Retrieved from https://gia.info.gov.hk/general/201506/17/P201506170559_0559_148059.pdf

Environment Bureau (2017a) *Number of Exceedances for SO2 in 2015 According to the WHO AQG*. Retrieved from https://gia.info.gov.hk/general/201703/22/P2017032200639_255568_1_1490180332433.pdf

Environment Bureau (2017b) *Hong Kong's Climate Action Plan 2030+*. Retrieved from www.climateready.gov.hk/files/report/en/HK_Climate_Action_Plan_2030+_booklet_En.pdf

Environmental Bureau (2019) *2025 Air Quality Objectives Review—Public Consultation*. Retrieved from www.gov.hk/en/residents/government/publication/consultation/docs/2019/Air-Quality.pdf

Environment Protection Department (2005) *Improving the Air Quality in Hong Kong – A Progress Report*. Retrieved from https://www.epd.gov.hk/epd/english/environmentinhk/air/prob_solutions/files/Brief_Progress_Report_Nov2005.pdf

Environment Protection Department (2015) *Air Quality in Hong Kong 2015*. Retrieved from http://news.cleartheair.org.hk/wp-content/uploads/2017/01/AQR2015e_final.pdf

Environment Protection Department (2016) *Environment Hong Kong 2016*. Retrieved from www.epd.gov.hk/epd/misc/ehk16/en/pdf1/web/EPD_AR_EHK2016.pdf

Environment Protection Department (2017) *LCQ11: Air Pollution in Hong Kong*. Retrieved from www.info.gov.hk/gia/general/201703/22/P2017032200639.htm?fontSize=1

Enviornmnetal Protection Department (2018) *PM2.5 Study for Air Quality Improvement in the Pearl River Delta Region—Feasibility Study*. Retrieved from www.epd.gov.hk/epd/sites/default/files/epd/english/environmentinhk/air/studyrpts/files/PM2.5_study_improve_prd_eng.pdf

Environmental Protection Department (2019a) *Air Quality Monitoring Network of Hong Kong*. Retrieved from www.aqhi.gov.hk/en/monitoring-network/air-quality-monitoring-network.html

Environmental Protection Department (2019b) *Hong Kong Air Pollutant Emission Inventory for 2017*. Retrieved from www.epd.gov.hk/epd/english/environmentinhk/air/data/emission_inve.html

Etzkowitz, H., & Leydesdorff, L. (1997) *Universities and the Global Knowledge Economy: A Triple Helix of University—Industry—Government Relations, (January 2001)*. SSRN.

Guangdong-Hong Kong-Macao Pearl River Delta Regional Air Quality Monitoring Network (2019) *Statistical Summary of the First Quarter Monitoring Results, (PRDAIR-2019–1)*. Retrieved from www.epd.gov.hk/epd/sites/default/files/epd/english/resources_pub/publications/files/PRD2019_1-en-1.pdf

Hao, J., Wang, S., Liu, B., & He, K. (2000) Designation of Acid Rain and SO_2 Control Zones and Control Policies in China. *Journal of Environmental Science and Health*, Part A, 35(10), 1901–1914.

Hasenfratz, D., Saukh, O., Walser, C., Hueglin, C., Fierz, M., & Thiele, L. (2015) Deriving High-resolution Urban Air Pollution Maps Using Mobile Sensor Nodes. *Pervasive and Mobile Computing*, 16, 268–285.

Helbich, M. (2018) Toward Dynamic Urban Environmental Exposure Assessments in Mental Health Research. *Environmental Research*, 161, 129–135.

HKSAR (2019a) *Press Release: LCQ22: Concentrations of Ozone in Air*. Retrieved from www.info.gov.hk/gia/general/201901/30/P2019013000782.htm

HKSAR (2019b) *Over 650 Datasets to be Released in First Year of Government Open Data Plans (with Photo/Video)*. Retrieved from www.info.gov.hk/gia/general/201901/03/P2019010300255.htm

HKSAR (2020a) *EPD Announces Air Quality Monitoring Results in 2019*. Retrieved from www.info.gov.hk/gia/general/202001/20/P2020012000874.htm

HKSAR (2020b) *28th Batch of Applications Approved under Pilot Green Transport Fund*. Retrieved from www.info.gov.hk/gia/general/202005/29/P2020052900236.htm

HKUST (n.d.) Evidence-based Ship Emissions Control Policies. *The Hong Kong University of Science and Technology*. Retrieved from https://ienv.ust.hk/ShipEmissions

HKUST, & Civic Exchange (2007) *Relative Significance of Local vs. Regional Sources: Hong Kong's Air Pollution*. Retrieved from www.hongkongcan.org/doclib/200703_RelativeSignificanceofLocalvsRegionalSources_Hong%20KongsAirPollution.pdf

Hora, R.M. (2010) Tackling Pollution at the Ports. *The Wall Street Journal*. Retrieved from www.wsj.com/articles/BL-HKB-1049

Hou, X., Chan, C.K., Dong, G.H., & Yim, S.H.L. (2019) Environmental Cooperation of Northeast Asia: Transboundary Air Pollution. *Environmental Research Letters*, 14.

HSN (2019) Hong Kong Air Pollution and the Deadly Impact of Shipping and Cruise Industries. *Hellenic Shipping News Worldwide (HSN)*. Retrieved from www.hellenicshippingnews.com/hong-kong-air-pollution-and-the-deadly-impact-of-shipping-and-cruise-industries/

Jansen, N. (2018) *Greater Bay Area: A Rebranding of the Pearl River Delta?* Retrieved from www.1421.consulting/2018/05/greater-bay-area/

Janssen, N.A.H., Gerlofs-Nijland, M.E., Lanki, T., Salonen, R.O., Cassee, F., Hoek, G., Fischer, P., Brunekreef, B., & Krzyzanowski, M. (2012) Health Effects of Black Carbon. *World Health Organization*. Retrieved from www.euro.who.int/__data/assets/pdf_file/0004/162535/e96541.pdf

Jin, Y., Andersson, J., & Zhang, S. (2016) Air Pollution Control Policies in China: A Retrospective and Prospects. *International Journal of Environmental Research and Public Health*, 13(12), 1219.

Kao, E. (2018, July 19) Government to Miss Roadside Pollution Targets as Hong Kong Levels Remain 70 Per Cent Higher than Recommended by World Health Organisation. *South China Morning Post.* Retrieved from www.scmp.com/news/hong-kong/health-environment/article/2155879/government-miss-roadside-pollution-targets-hong

Keegan, M. (2018) Shenzhen's Silent Revolution: World's First Fully Electric Bus Fleet Quietens Chinese Megacity. *The Guardian.* Retrieved from www.theguardian.com/cities/2018/dec/12/silence-shenzhen-world-first-electric-bus-fleet

Khreis, H., Ramani, T., de Hoogh, K., Mueller, N., Rojas-Rueda, D., Zietsman, J., & Nieuwenhuijsen, M.J. (2019) Traffic-related Air Pollution and the Local Burden of Childhood Asthma in Bradford, UK. *International Journal of Transportation Science and Technology*, 8(2), 116–128.

Kim, I. (2007) Environmental Cooperation of Northeast Asia: Transboundary Air Pollution. *International Relation of the Asia-Pacific*, 7, 439–462.

Larssen, T., Lydersen, E., Tang, D., He, Y., Gao, J., & Liu, H. (2006) Acid Rain in China. *Environmental Science Technology*, 40(2), 418–425.

Lau, K.H., Yu, J.Z., Wong, T.W., Yu, T.S., & Moore, M. (2004). *Assessment of Toxic Air Pollutants in Hong Kong.* Technical Report submitted to the Hong Kong Environmental Protection Department.

Lee, C.H., Wang, Y.B., & Yu, H.L. (2019) An Efficient Spatiotemporal Data Calibration Approach for the Low-cost PM2.5 Sensing Network: A Case Study in Taiwan. *Environmental International*, 130, 104838.

Lee, M., Brauer, M., Wong, P.Y.P., Tang, R., Tsui, T.H., Choi, C., Cheng, W., Lai, P.C., Tian, L.W., Thach, T.Q., Allen, R., & Barratt, B. (2017) Land Use Regression Modelling of Air Pollution in High Density High Rise Cities: A Case Study in Hong Kong. *Science of the Total Environment*, 592, 306–315.

Li, H., Wang, J., Li, R., & Lu, H. (2018) Novel Analysis—forecast System Based on Multi-objective Optimization for Air Quality Index. *Journal of Cleaner Production*, 208, 1365–1383.

Li, Y., Lin, C., Lau, A.K.H., Liao, C., Zhang, Y., Zeng, W., Li, C., Fung, J.C.H., & Tse, T.K.T. (2015) Assessing Long-Term Trend of Particulate Matter Pollution in the Pearl River Delta Region Using Satellite Remote Sensing. *Environmental Science & Technology*, 49(19), 11670–11678.

Liang, X., Li, S., Zhang, S., Huang, H., & Chen, S.X. (2016) PM2.5 Data Reliability, Consistency, and Air Quality Assessment in Five Chinese Cities. *Journal of Geophysical Research: Atmospheres*, 121(17), 10220–10236.

Lin, C., Li, Y., Lau, A.K.H., Li, C., & Fung, J.C.H. (2018) 15-Year PM2.5 Trends in the Peal River Delta Region and Hong Kong from Satellite Observation. *Aerosol and Air Quality Research*, 18, 2355–2362.

Lin, M., Chan, L.N., Chan, C.Y., Wang, X.M., & Dong, H.Y. (2011) *Emerging Air Pollution Issues in Changing Pearl River Delta of South China by The Impact of Air Pollution on Health, Economy, Environment and Agricultural Sources*, Khallaf, M.K. (ed.). Retrieved from www.intechopen.com/books/the-impact-of-air-pollution-on-health-economy-environment-and-agricultural-sources/emerging-air-pollution-issues-in-changing-pearl-river-delta-of-south-china

Liu, S., Xing, J., Wang, S., Ding, D., Chen, L., & Hao, J. (2020) Revealing the Impacts of Transboundary Pollution on PM2.5-related Deaths in China. *Environment International*, 134, 105323.

Lu, X.C., Yao, T., Li, Y., Fung, J.C.H., & Lau, A.H. (2016) Source Apportionment and Health Effect of NOx Over the Pearl River Delta Region in Southern China. *Environmental Pollution*, 212, 135–146.

Lu, X.C., Zhang, S., Xing, J., Wang, Y., Chen, W., Ding, D., Wu, Y., Wang, S., Duan, L., & Hao, J. (2020) Progress of Air Pollution Control in China and Its Challenges and Opportunities in the Ecological Civilization Era. *Engineering* (In Press).

MEP (Ministry of Environmental Protection) (2010) Facilitating the Joint Prevention and Control of Air Pollution and Improving the Regional Air Quality. *MEP, the People's Republic of China*. Retrieved from http://english.mep.gov.cn/News_service/news_release/201010/t20101025_196606.htm

Miskell, G., Salmondb, J., & Williams, D.E. (2017) Low-cost Sensors and Crowd-sourced Data: Observations of Siting Impacts on a Network of Air-quality Instruments. *Science of the Total Environment*, 575, 1119–1129.

Moltchanov, S., Levy, I., Etzion, Y., Lerner, U., Broday, D.M., & Fishbain, B. (2015) On the Feasibility of Measuring Urban Air Pollution Bywireless Distributed Sensor Networks. *Science of the Total Environment*, 502, 537–547.

Ng, S. (2018) Fair Winds Charter: How Civic Exchange Influenced Policymaking to Reduce Ship Emissions in Hong Kong 2006–2015. *Civic Exchange*. Retrieved from https://civic-exchange.org/wp-content/uploads/2018/09/FairWinds Charter_2018_REPORT.pdf

Nip, A. (2017, November 21) Call for More Coverage Over Air Quality. *The Standard*. Retrieved from www.thestandard.com.hk/section-news.php?id=189847&story_id=47422012&d_str=20171121&sid=4

Schneider, P., Castell, N., Vogt, M., Dauge, F.R., Lahoz, W.A., & Vartonova, A. (2017) Mapping Urban Air Quality in Near Real-time Using Observations from Low-cost Sensors and Model Information. *Environmental International*, 106, 234–247.

Shi, Y., Xie, X., Fung, J.C.H., & Ng, E. (2018) Identifying Critical Building Morphological Design Factors of Street-level Air Pollution Dispersion in High-density Built Environment Using Mobile Monitoring. *Building and Environment*, 128, 248–259.

StateAir (n.d.) *Mission in China, U.S. Department of State Air Quality Monitoring Program*. Retrieved from www.stateair.net

Stoerk, T. (2016) Statistical Corruption in Beijing's Air Quality Data Has Likely Ended in 2012. *Atmospheric Environment*, 127, 365–371.

Sudiana, K., Sule, E.T., Soemaryani, I., & Yunizar, Y. (2020) The Development and Validation of the Penta Helix Construct. *Business: Theory and Practice*, 21(1), 136–145.

Sunyer, J., Suades Gonzalez, E., García-Esteban, R., Rivas, I., Pujol, J., Alvarez-Pedrerol, M., Forns, J., Querol, X., & Basagana, X. (2017) Traffic-related Air Pollution and Attention in Primary School Children. *Epidemiology*, 28, 181–189.

Thach, T. Q., Tsang, H., Cao, P., & Ho, L. M. (2018) A Novel Method to Construct an Air Quality Index Based on Air Pollution Profiles. *Int J Hyg Environ Health*. 221(1), 17–26.

Trade Development Council (2020) *Statistics of the Guangdong-Hong Kong-Macao Bay Area*. Retrieved from https://research.hktdc.com/en/article/MzYzMDE5NzQ5

Transport and Housing Bureau (2018) *Summary Statistics on Port Traffic of Hong Kong*. Retrieved from www.hkmpb.gov.hk/document/summary_statistics_22feb.pdf

Tzivian, L., Winkler, A., Dlugaj, M., Schikowski, T., Vossoughi, M., Fuks, K., & Hoffmann, B. (2015) Effect of Long-term Outdoor Air Pollution and Noise on Cognitive and Psychological Functions in Adults. *International Journal of Hygiene and Environmental Health*, 218(1), 1–11.

Uno, I., Osada, K., Yumimoto, K., Wang, Z., Itahashi, S., Pan, X., & Nishizawa, T. (2017) Importance of Long-range Nitrate Transport Based on Long-term Observation and Modeling of Dust and Pollutants Over East Asia. *Aerosol and Air Quality Research*, 17(12), 3052–3064.

Voiland, A. (2018) *Air Pollution Shrouds the Pearl River Delta, NASA Earth Observatory*. Retrieved from https://earthobservatory.nasa.gov/images/91614/air-pollution-shrouds-the-pearl-river-delta

Wang, X.M., Lin, W.S., Yang, L.M., Deng, R.R., & Lin, H. (2007) A Numerical Study of Influences of Urban Land-use Change on Ozone Distribution Over the Pearl River Delta Region, China. *Tellus B*, 59(3), 633.

Wong, C.M., Lam, T.H., Peters, J., Hedley, A.J., Ong, S.G., Tam, A.Y., Liu, J., & Spiegelhalter (1998) Comparison between Two Districts of the Effects of an Air Pollution Intervention on Bronchial Responsiveness in Primary School Children in Hong Kong. *Journal of Epidemiology & Community Health*, 52, 571–578.

Wong, C.M., Tsang, H., Lai, H.K., Thomas, G.N., Lam, K.B., Chan, K.P., Zheng, Q., Ayres, J.G., Lee, S.Y., Lam, T.H., & Thach, T.Q. (2016) Cancer Mortality Risks from Long-term Exposure to Ambient Fine Particle. *Cancer Epidemiology Biomarkers Prevention*, 25(5), 839.

Wong, N.S., Leung, C.C., Li, Y., Poon, C.M., Yao, S., Wong, E.L.Y., Lin, C., Lau, A.K.H., & Lee, S.S. (2017) PM2.5 Concentration and Elderly Tuberculosis: Analysis of Spatial and Temporal Associations. *Lancet*, 390, S68.

Wong, P.P.Y., Lai, P.C., Allen, R., Wei, C., Lee, M., Tsui, A., Tang, R., Thach, T.Q., Tian, L., Brauer, M., & Barratt, B. (2019) Vertical Monitoring of Traffic-related Air Pollution (TRAP) in Urban Street Canyons of Hong Kong. *Science of the Total Environment*, 670, 696–703.

Wong, T.W., Tam, W.S., Yu, T.S., & Wong, A.H.S. (2002) Associations between Daily Mortalities from Respiratory and Cardiovascular Diseases and Air Pollution in Hong Kong, China. *Occupational and Environmental Medicine*, 59, 30–35.

World Health Organization (2018) *Ambient (Outdoor) Air Pollution*. Retrieved from www.who.int/en/news-room/fact-sheets/detail/ambient-(outdoor)-air-quality-and-health

Wu, D., Wu, X., Li, F., Tan, H., Chen, J., & Cao, J. (2010) Temporal and Spatial Variation of Haze During 1951–2005 in Chinese Mainland. *Acta Meteorol Sinica*, 68(5), 680–688.

Yang, L., Stulen, I., De Kok, L.J., & Zheng, Y. (2002) SO2, NOx and Acid Deposition Problems in China—Impact on Agriculture. *Phyton* (Austria) Special Issue: "Global Change", 42(3), 255–264.

Yang, Y., Tang, R., Qiu, H., Lai, P.C., Wong, P., Thach, T.Q., Allen, R., Brauer, M., Tian, L., & Barratt, B. (2018) Long Term Exposure to Air Pollution and Mortality in an Elderly Cohort in Hong Kong. *Environment International*, 117, 99–106.

Zhang, M., & McSaveney, M.J. (2018) Is Air Pollution Causing Landslides in China? *Earth and Planetary Science Letters*, 481, 284–289.

Zhong, L., Louie, P.K.K., Zheng, J., Yuan, Z., Yue, D., Ho, J.W.K., & Lau, A.K.H. (2013) Science Policy Interplay: Air Quality Management in the Pearl River Delta Region and Hong Kong. *Atmosphere Environment*, 76, 3–10.

Index

Note: Page numbers in *italics* indicate a figure and page numbers in **bold** indicate a table on the corresponding page. Page numbers followed by "n" indicate a note.

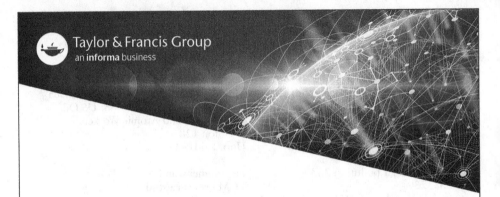

Printed in the United States
by Baker & Taylor Publisher Services

Printed in the United States
by Baker & Taylor Publisher Services